THE LAST 10,000 YEARS

A Fossil Pollen Record
of the
American Southwest

Paul S. Martin

THE UNIVERSITY OF ARIZONA PRESS
TUCSON   1963

The University of Arizona Press
www.uapress.arizona.edu

Printed in the United States of America
21   20   19   18   17   16   7   6   5   4   3   2

ISBN-13: 978-0-8165-0050-5 (cloth)
ISBN-13: 978-0-8165-3535-4 (Century Collection paper)

L. C. Catalog Card No. 63-11984

♾ This paper meets the requirements of ANSI/NISO Z39.48-1992 (Permanence of Paper).

CONTENTS

ILLUSTRATIONS

TABLES

FOREWORD

The name of palynology is fairly new, but the subject is ancient, and may be older in Arizona than anywhere else in the world. *Palyn-,* cognate with *pollen,* is literally dust — not the kind that settles on old books and scholars, but the vital, masculine dust that quickens ovules, and is ceremonially blown to the six directions by Pueblo shamans. The dust of palynology is partly made up of spores of ferns and fungi, and is therefore not all pollen; botanists, who coined the term, prize the distinction, but the difference, like gold-plating on the ripcord of a parachute, is not one that matters once the object is airborne. It matters even less to the author of this book, who is concerned with fossilized husks of pollen grains and spores. For him, the airborne dust of ages has settled in layers, one layer for each age, which he can dissect. As "time . . . hath an art to make dust of all things," the layers turn out to be mostly dross, but the pollen panned from them is a matchless pile of documents. The history so extracted is vegetable in character, as might be expected, but it refers, more than incidentally, to four hundred generations of mankind in Arizona.

Since the time of the German geologist Ehrenberg, a hundred years ago, peat bogs and lake muds in northern Europe have been known to contain records of this sort. As long ago as 1916 von Post showed how to reconstruct ancient vegetation, by counting the kinds of pollen it shed, level by level, in the growing peat bogs of Sweden. To describe a kind of vegetation is to specify a kind of climate, or of land use; once a sequence of these is known, any of its stages can be read as a dated event, like a piece of Kleenex in a heap of Aztec potsherds. So von Post's simple trick of counting pollen has been used, in the boggier parts of the world, to weave a rich tapestry of historical fact. Its fabric is a chronicle of climatic change, which geophysicists and oceanographers are finding increasingly fascinating; among the distinctly nonbotanical events that are worked in are the explosion of Mount Mazama to form Crater Lake and the flooding of villages of the Swiss lake-dwellers.

What Paul Martin and his colleagues have done, then, is not to invent a new methodology, but to apply an old one in a new context. The American Southwest is a notably unboggy country, and lakes, which serve as well as bogs as traps for airborne pollen, are not as common as they used to be. Shortage of water has long been supposed to mean a shortage of mud, of the pollen-bearing kind from which history is read. For the same reasons the Mediterranean lands, so rich in other sorts of historic dust, have not been much prospected for pollen. It happens, though, that the dry lakes of Arizona, if one digs under the mirage, contain a good deal of pollen, and the streams have also carried a lot during their intermittent lives. With patience and much technical skill the author has separated and counted it, and the history that comes out is quite new at many points. For instance, the time of most severe erosion, in the last ten thousand years, was probably moister, not drier, than the present; the period of the Sulphur Spring people, circa 6000 B.C., was not especially rainy and certainly not part of a glacial age; Arizona's big game roamed the desert plains, successfully if not happily, for thousands of years after the long rains ended, and then died out, about 5000 B.C., for reasons that evidently had little to do with climate.

Impressive as the achievement is, many problems remain — some familiar, and some new, exposed by the breaking of so much new ground. That business of blowing pollen around in Pueblo ceremonies, for example. Ancient people are permitted — encouraged, in fact — to interfere with vegetation, but to interfere with the pollen is to play pretty fast and loose with history; piety, famous for falsifying documents, has rarely gone to such lengths. We know it was done by the prehistoric Hopi; luckily, the practice seems not to have been widespread. Among more conventional problems, there is the ambiguity inherent in reconstructing vegetation from pollen. If all plants produced the same amount of pollen, if wind and water carried it an equal distance, and if all kinds could be surely recognized, palynology might be an exact science. Then, too, if the layers of silt in arroyos were like layers of peat in bogs, the prehistory of Arizona would be less a thing of shreds and patches, fitted together neatly but not without voids. Without such problems, though, the study would not have been worth doing; there would have been no scope for historical insight, of which the author has plenty. His pioneering work is exciting, not only for what it tells, but even more for what it promises: history, free of peat-bound preconceptions, in a land of little mud.

Yale University   Edward S. Deevey, Jr.
March, 1963

INTRODUCTION

During a summer in the Chiricahua Mountains of southeastern Arizona in 1956, I became acquainted with some of the montane Mexican animals and plants which barely cross the international boundary and which are presently isolated in subhumid habitats of the Southwestern mountains. Mexican species of fence lizards, rattlesnakes, deer mice, tree squirrels, bats, and many species of plants including the characteristic montane Mexican pines and oaks have their centers of distribution south of the Chiricahua Mountains and appear to be dependent on the distinctive summer rain climate of the Mexican Plateau. The Mexican species isolated in parts of southern Arizona and New Mexico are scarcely differentiated from related populations to the south. Although fossil evidence was scanty, it did not seem possible that the Mexican animals and plants could have survived in the Southwest during the glacial period 20,000 years ago when, under a pluvial climate, the Southwest was being invaded by Cordilleran biotas from the north.

To explore climatic history further I began a study of the late Pleistocene fossil pollen record in the vicinity of the Chiricahua Mountains, hoping in this way to establish the most recent period of contact between pine-oak woodland biotas of the Sierra Madre Occidental with those isolated in mountain ranges of the Southwest. In the process of pursuing this rather limited biogeographic objective it became apparent that its scope might be widened. What was the climatic sequence of the last 10,000 years in the Southwest? Was it marked by intensive drought as has been thought by so many? What does the pollen stratigraphic record reveal about extinction of late Pleistocene mammoths and other large mammals whose disappearance commonly defines the end of the Pleistocene in the mind of the geologist? Can fossil pollen in the Southwest be used, as it is in the Old World, to guide archaeological excavations, to reveal the rise of prehistoric agriculture, and to record Pleistocene biotic change? Immediately, questions of method arise. How might the pollen content of arid land sediments best be used to study prehistory and climatic change? Can it be used with confidence?

The scope of inquiry has outrun my initial objective and while it appears that the fossil pollen record gives some valuable clues about the time of arrival of the Mexican woodland biota, that problem is far from settled. But it is clear that the pollen record of southern Arizona 20,000 years ago is so different from anything known in the last 10,000 years that we can speak with confidence about the elimination of Mexican grassland and woodland habitats from the vicinity of the Chiricahua Mountains in the Pleistocene and put 15,000 years as an approximate limit on their return. The present landscape and vegetation pattern has developed since then.

To uncover the postluvial pollen record — the last 10,000 years — I have chosen the alluvium of the inner valley flood plains, for the following reason. Pollen analysis in the Southwest began with Paul Sears' analysis (157) of alluvial silts collected by Ernest Antevs near Kayenta, Arizona. More recent pollen studies have centered mainly on buried lake sediments including those of dry lakes or playas (12, 34, 35, 74, 75, 84, 147, 158) and caves (122, 159). The lacustrine pollen record reveals major changes in climate of the Southwest — changes involving a Wisconsin-age drop of vegetation zones by 1,200 meters.

While the lake and playa pollen records help a great deal to reveal the vegetation and climate during and before the Wisconsin pluvial period, they leave a gap in our knowledge of more recent events. The playa lake beds would not be expected to yield a sedimentary record when they were dry or seasonal, as was the case during most of postpluvial time. Permanent lakes are rare in arid regions and except under unusual circumstances as at Montezuma Well the lakes are confined to high elevations. Strangely, even some of the high elevation permanent lakes lack a rich sedimentary record of the last 10,000 years; they contain mainly pluvial-age pollen (12). Caves are promising sources of a postpluvial record but they require costly and very careful excavation. To reconstruct the postpluvial climatic history of the Southwest one must return to the ubiquitous flood plain deposits of the type first examined by Sears.

Flood plains have an ecology of their own. The Southwestern flood plains provided the main source of water to man and animals in postpluvial times. They also provided highly productive natural pastures attractive to large herbivores. Surface water and forage along the flood plains was essential to the 19th century trappers, military parties, and settlers traveling to California by horse, mule, and wagon. Flood plain soils and water provided the principal environment for farming in prehistoric as well as in modern times. If the crops have shifted from production of corn, beans, and squash by the Hohokam in the 12th century to cotton, alfalfa, and lettuce of the modern farm operator, and if the method of irrigation has shifted from the use of natural runoff to pumping of ground water from great depths, the basic soil requirement for a successful harvest in the Southwest remains the same — fine alluvium of the valley bottoms, well watered.

The prehistory of Southwestern flood plains is found in their eroded and exposed beds which contain a fossil record extending back thousands of years, occasionally to the end of the last pluvial period, 10,000 to 11,000 years ago. Buried in them are the remains of mammoths, camels, tapirs, and other extinct animals of the late Pleistocene. Here also have been found both the projectile points of the last big game hunters and the simple grinding tools of the first plant harvesting people in the Southwest. Subsequent cultural history can be inferred from the appearance of more elaborate grinding tools such as manos and basin metates, and in the development of pit houses and storage pits constructed by later inhabitants. A notable event in the record of flood plain artifacts is the appearance of pottery about 2,000 years ago. Then, shortly before arrival of Western Man in the 16th century, there is an intriguing and ecologically unexplained decline of the flood plain Indian tribes through many parts of the Southwest. Finally, in the late 19th century, occurred the last episode of arroyo cutting that excavated the present channels and exposed buried sediments containing the fossil history of the last 10,000 years. For all of these reasons, the Southwestern flood plains contain the key to answers long sought by many interested in prehistory, climatic change, and biogeography.

Admittedly, this work is not definitive. Certain of the findings point to the need for a wider and deeper study, and a more critical approach. I expect criticisms of methodology and of interpretations; they are welcome. If the evidence I have assembled leads to a re-examination of previously accepted hypotheses about the environment and life of the Southwest during the last 10,000 years, I believe the result will be a closer unity of the several great disciplines whose particular interests center on understanding the prehistoric period. While interdisciplinary cooperation is highly desirable it is not an excuse for passive accommodation to fashionable theories, or for an uncritical approach.

Although the flood plain palynologist has the good fortune of being able to relate his pollen record to the most important stratified prehistoric sites known in the Southwest, interpretation of the flood plain pollen record is another matter. Thorough ecological studies of many of the important pollen-producing plants of flood plains are virtually non-existent. Furthermore, it should be obvious to the most casual student of geology that episodes of erosion will result in a broken stratigraphic record, generally of unknown duration. Some pollen stratigraphers shun alluvium entirely, feeling that the likelihood of erosion and redeposition of pollen from older alluvial fill is so great that pollen counts of alluvium cannot be used with confidence. I have adopted a pragmatic view, feeling that alluvium is worthy of study even if a record of redeposition is the main result. As will be seen, the concern about the danger of redeposition remains more theoretical than real and the burden of proof must now be assumed by those who claim that redeposition is a serious problem.

While much remains to be discovered about pollen cycling in alluvium, redeposition does not obliterate a consistent stratigraphic pattern which can be correlated between adjacent valleys and which, on the basis of radiocarbon dates, appears to involve deposition of the same pollen types at approximately the same time throughout the desert grassland of the Southwest. No other terrestrial fossils are so widely distributed or so closely related to past vegetation and climate. If this is so, it appears reasonable to assume that we can refine our notions about flood plains based on paleontology, geology, archaeology, and tree-ring analysis.

With pleasure I recall the generosity of Ernst Antevs, Charles C. DiPeso, Frank Eddy, Emil W. Haury, Edwin B. Kurtz, John Lance, Mark A. Melton, Edward B. Sayles, and Terah L. Smiley in either directing me to suitable localities or collecting pollen samples or sharing their knowledge of Southwestern prehistory. In the matter of pollen identification, Jane Gray and Lucy Cranwell Smith came to the rescue on various occasions. James E. McDonald and William D. Sellers gave freely of their climatological experience. Richard H. Hevly, Robert R. Humphrey, S. Clark Martin, Charles T. Mason, Raymond M. Turner, David G. Wilson, and Tien Wei Yang helped in matters of plant distribution and identification. Charles H. Lowe, Joe T. Marshall, and Robert H. Whittaker provided information regarding regional ecology and biogeography. Marvin A. Stokes photographed certain habitats. A special acknowledgment is reserved for James Schoenwetter who analyzed half of the profiles and for Bernard C. Arms who carried out all the extractions.

Finally, it may be appropriate to recall that Pleistocene paleoecology in the Southwest stems largely from the contributions of Kirk Bryan and Ernst Antevs. If results of the present investigation lead to a revision of their conclusions, it is a further testimony to their skill as scientific pioneers.

Research on Southwestern pollen stratigraphy was made possible through grants from the National Science Foundation and the American Philosophical Society.

Tucson 1963   Paul S. Martin

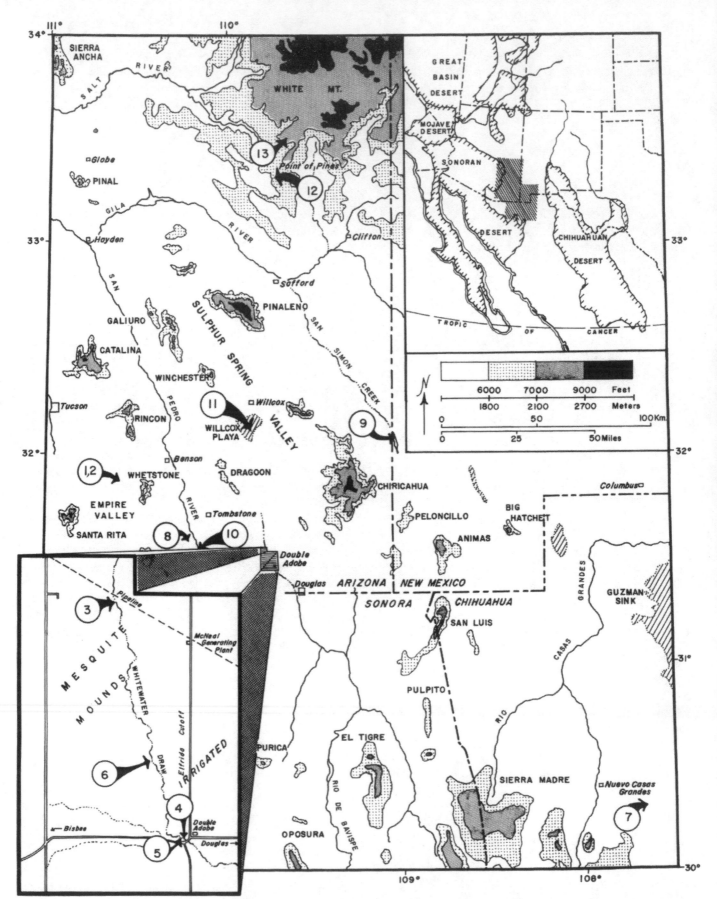

Figure 1. Location of study areas: 1. Cienega Creek, Empire Valley; 2. Matty Wash, Empire Valley; 3. Double Adobe IV; 4. Double Adobe I; 5. Double Adobe II; 6. Double Adobe III; 7. Malpais Site; 8. Murray Springs; 9. San Simon Cienega; 10. Lehner Site; 11. Willcox Playa; 12. Cienega Creek Site, Point of Pines; 13. Dry Prong Reservoir.

I. GEOLOGIC FRAMEWORK

South of the Mogollon Rim of the Colorado Plateau and north of the Sierra Madre Occidental lies the basin and range physiographic province. The landscape features relatively small, high, conspicuous, isolated mountain ranges with a northwest trend (Fig. 1). The available relief commonly exceeds 1,000 m. and may exceed 2,000 m. Most of the study area lies west of the Continental Divide in the neighborhood of the Chiracahua Mountains in desert grassland at 1,200 m.–1,350 m. elevation.

In southern Arizona the basin and range province is predominantly of through-flowing (although intermittent) drainage. The only large closed basin in southeastern Arizona is the Willcox Playa or Playa de los Pimas. Such basins are larger and more numerous to the southeast in Chihuahua and to the northwest in California and Nevada. Small closed basins occur in craters of volcanic fields near Pinacate, Sonora and San Bernardino, Arizona. Natural lakes are unknown in the basin-range mountains of Arizona.

Bajadas. The Southwestern landscape is characterized by long, gentle slopes utterly unlike any landform familiar to those who live outside the desert. Close to the mountains the slopes or bajadas overlie a shallow bedrock platform. At a greater distance from the mountains the slopes overlie deep valley fill, generally of conglomerate, sometimes lacustrine clays, often intruded by volcanic flows, the whole extending to a depth of several thousand feet. In the San Simon Valley of Cochise County the valley fill is 2,150 m. deep, reaching well below sea level (150). In the lowest point of the valley adjacent bajadas meet at a flood plain or, if the basin is closed, at a playa. Adjacent bajadas may differ in color and in vegetation if the mountain ranges providing the rock waste are not of the same lithology.

Following local Spanish usage, Hill (85) adopted the term bajada to apply to the gentle valley slopes, whether eroding or aggrading, whether underlain by bedrock or by fill. As Hill stressed, bajadas are not structural features. Subsequent generations of Southwestern botanists, zoologists, and geographers have generally followed Hill's usage. On the other hand, geologists commonly restrict the term bajada to undissected, coalescing fans and refer to the eroding slopes as pediments. Whatever one calls the gentle intermontane slopes, they characterize much of the landscape of arid America, they function as important surfaces of transport in erosion of the desert mountains, and they support plant communities distinct from those in the mountains above and in the mid-valley flood plains or playas below.

Biologists commonly divide Southwestern bajadas into two distinct geomorphological and ecological units. The upper bajada usually overlies bedrock, often mantled by a well-drained gravelly soil, and harbors a varied plant community. The lower bajada typically overlies deep fill, is covered by a fine-grained and highly calichified soil, and harbors a less varied plant community (see Fig. 6).

The bajada is seen to best advantage in the Mojave Desert of California and Nevada, in the Chihuahuan Desert of Mexico, and in southeastern Arizona. It is inconspicuous in the southern part of the Sonoran Desert south of Hermosillo, where the ecological distinction of upper and lower bajada vegetation vanishes (169).

At present almost all bajadas of southeastern Arizona are being dissected, particularly along the San Pedro Valley near Benson where three distinct erosional surfaces or pediments are evident (68). In the Sulphur Spring Valley a weathered, rounded lag gravel on the Mule Mountain bajada along the High Lonesome road 15 km. west of McNeal represents an ancient river deposit. According to Mark Melton (pers. corr.) it indicates at least 120 m. of stripping in the Sulphur Spring Valley, the vertical distance between the gravel and the elevation of the Whitewater Draw. A history of extensive erosion might not be expected in view of the evenly graded, undissected appearance of the Mule and Swisshelm bajadas.

A broad platform of bedrock extending beyond the mountain backwall and underlying the upper bajada is commonly exposed in arroyo cuts. Current studies lead to the conclusion that the planed bedrock platform or rock pediment is an exhumed Miocene surface, perhaps formerly covered by alluvium of unknown depth (174, 100).

Valley Fill. A perennial problem in Cenozoic geology of the Southwest is the origin, age, correlation, and climatic history of the valley fill. Can these important deposits, some of which bear a superficial resemblance to glacial drift, be understood in terms of Pleistocene climatic change? Huntington (92) thought so, theoriz-

1

ing that deposition occurred in interglacial time and erosion in glacial periods. Before pursuing this further I will summarize briefly more recent findings in the study of valley fill.

Considering the total bajada *surface* from mountain backwall to inner valley alluvium, one commonly observes a gradation in sediment size. Large boulders and coarse gravel occur on the upper bajada adjacent to the mountains, while finer material is carried out toward the center of the valley. Traditionally it has been assumed that a similar gradient also typifies the buried fill. Gravel near the mountains and silt in the middle of the valley is what one might expect if the main source of valley fill were from erosion of the mountain backwall.

Despite the expectation, Lance (99, 100) and Melton (133) have pointed out that in deeply dissected valleys *fine-grained sediment without boulders or gravel* may extend to the margins of the valleys and may lap against the mountains, as at the 111 Ranch near Safford. In the Sulphur Spring Valley lacustrine gypsum and lake-marl deposits, considered of Pleistocene age, lie against the Perilla Mountains (36). Such field evidence poses a new problem regarding the origin of valley fill. The fine-grained sediments adjacent to the mountain backwall in the basin-range of southern Arizona may not be of local origin. Melton (133) postulates that some sediments were carried in by preorogenic rivers flowing before the mountains existed in their present form.

In his dissertation Heindl (81) recognized two major types of valley fill, a "lower set" and an "upper set." The "lower set" includes those deposits ". . . that (1) are apparently not directly related to the major structural troughs forming the basins of the existing Basin and Range topography; (2) are at least locally deformed by thrust and normal faulting; (3) may be intruded and mineralized." The "upper set" deposits are characterized by ". . . (1) lithology which reflects the rocks in the adjacent mountain areas; (2) sedimentary relationships which reflect deposition essentially within the topographic basins in existence today; (3) structural relationships involving the development of basins and ranges along normal faults; (4) lack of mineralization; and (5) erosion to approximately the same degree of relief." The "lower set" can be shown to be older than "upper set" deposits when the two are in contact. Both have been called Gila Conglomerate, a name which Heindl rejects.

Regarding age, Heindl (81) considered that most of the valley fill in the structural troughs of southeastern Arizona and southwestern New Mexico ranges from at least early Miocene to early Pleistocene, probably Kansan. A recent potassium-argon date of 34.8 ± 2.1 million years on an ash flow near Davidson Canyon in Pima County indicates an early Oligocene age of the Pantano beds which are part of Heindl's "lower set" of the valley fill (41). According to Lance (100)

". . . the bulk of the valley fill appears to represent deposition extending through a large part of Pliocene time and continuing into middle Pleistocene . . . All late Pleistocene fossil localities listed are either from caves or occur in superficial deposits."

Climates during deposition of lacustrine beds in valley fill near Safford were more or less humid and warm (98). Seventy-three percent of the small rodents reported in the Blancan-age beds near the town of Benson are presently found at the foot of the Huachuca Mountains (86). The elevation of Fort Huachuca is about 400 m. above that of the fossil locality; it receives 150 mm. more rainfall. If the rodent fauna is a suitable index, the climate during deposition of the Benson beds was appreciably less arid than at present.

The 111 Ranch fauna near Safford has recently yielded microtine rodents and the pocket gopher *Geomys* (183). Microtines do not live near Safford today and their presence would represent a vertical displacement of at least 600 m.; thus a cooler climate may be inferred from the fauna.

In the hope that the pollen record would provide additional clues to past climate during deposition of the valley fill, a variety of samples was examined. Initial results were disappointing. It appears that intense weathering of exposed beds may have removed all but poorly preserved remnants of the former pollen content (73). Subsurface samples, especially those below the water table, are much more promising. A core collected by the Kennecott Copper Company near Safford, Arizona, depth 257-262 ft., and analyzed by Gray, contained well-preserved pine (52.6 percent), sagebrush (11.0 percent), fir (7 grains), *Sarcobatus* (3.6 percent), and other indicators of a major change in climate. A second core from the same region revealed 16 to 27 percent juniper-cypress, five to 10 percent pine, and 13 percent sagebrush (74). Gray's spectra are the first early Pleistocene (possibly Nebraskan-age) pollen floras reported from the arid Southwest. They reveal more moist and probably cooler climates at the eastern edge of the Sonoran Desert and confirm the geological record of more moisture during episodes of valley filling.

Why filling did not continue into the late Pleistocene is unclear; it is possible that certain late Pleistocene fill remains to be discovered; if so, it is definitely not of the thickness of that known from the early and middle Pleistocene (Blancan and Irvingtonian faunas). Huntington's simple model of deposition and erosion of fill under Pleistocene climatic control can be dismissed.

Inner-valley Alluvium. A major feature in Pleistocene geology of Arizona is the substantial time gap, presumably of hundreds of thousands of years, which separates the postglacial pollen-bearing inner-valley alluvium from the underlying deposits of fill. The inner-valley alluvium is a flood plain deposit varying from a few hundred meters to several kilometers in width, depending on extent of the drainage system. Along the Gila it exceeds 30 m. in depth, but along most drainages it is much shallower. Near Double Adobe relative depth of inner-valley alluvium (black) and valley fill (stippled) is shown on Fig. 6.

The inner-valley alluvium may contain cobbles, as in the lower portion of the Malpais site (Plate 14); it may contain crossbedded sand, as at Double Adobe I (Plate 13); it may be largely of sand, silt, and clay as along the Cienega Creek near Matty Canyon. Dark bands relatively rich in organic materials which show to advantage in banks of the San Pedro near Benson and at Matty Canyon (Plate 11) are local features which mark the position of former cienegas (wet meadows). The pollen record of the dark bands shows that despite their humified condition they do not represent past periods of heavy rainfall.

Prior to channel cutting of the late 19th century, extensive cienegas and shallow semipermanent streams characterized many of the flood plains in southern Arizona (22, 23). Ground water lay close to the surface through the year and a dense plant cover of phreatophytes included sacaton grass (*Sporobolus wrightii*), with sedges in wet pockets and occasionally tule (*Scirpus*) or cat-tail (*Typha*). After erosion the cienegas virtually disappeared. Remnants persist along the Empire Wash in southern Pima County (Plate 2), near the headwater of the San Pedro River in northern Sonora, and at San Bernardino Ranch east of Douglas. The San Simon Cienega north of Rodeo along the Arizona-New Mexico state line is perhaps the largest remaining example of this intriguing Southwestern habitat.

Because of its cultural content the inner-valley alluvium of Southwestern flood plains has attracted considerable interest. Three major periods of cutting and three of filling are commonly encountered. Various authors have presented regional correlations of the cut-fill sequence (24, 25, 95, 136). Table 8 is adapted from Miller (135). Local minor variants are reported; for example, in some areas deposition phase II appears to have two subdivisions. While cutting is still occurring in most flood plains, channel filling has begun in others (Photograph 9).

While it is commonly thought that the major episodes of deposition and cutting throughout the South-west are synchronous rather than random, this assumption deserves further testing. The fossil pollen record should reveal biotic and climatic conditions during deposition, but it may not provide continuous evidence of conditions during the time of cutting.

The period of headwater arroyo cutting today occurs mainly in the summer. This is the season of convectional storms and flash floods. In studies conducted near Tombstone and elsewhere in Arizona and New Mexico the Southwest Watershed Hydrology Group found that 97 percent of the summer storms have a diameter of less than 3.0 miles; 80 percent of these storms lasted less than 4.0 hours, and 80 percent reached peak intensities of precipitation greater than two inches per hour (97). *For a period of a few minutes* precipitation intensity may exceed 10 inches per hour, and a record intensity, 24.5 inches per hour, was recorded at Alamogordo Creek watershed in the summer of 1960. Within minutes after the storm begins dry arroyos receive a peak discharge of as much as 6,000 cubic feet per second. Two hours later a scoured channel and a trickle of water may be all that is left to mark the storm's passage.

In contrast, the frontal storms of winter provide much lower rainfall intensity; runoff gradually increases downstream, but the headwater tributaries experience no or very little channel flow.

The findings of the Southwest Hydrology Group indicate that both runoff and transmission losses in the channel are at a maximum after the first summer rain. Until the channel is thoroughly moist and highly permeable beds are filled, extensive channel routing may not occur. Wet initial channel conditions increase rate and volume of channel flow. Such conditions would occur in especially wet summers when repeated heavy rains should bring about "arroyo cutting."

Assuming a similar precipitation pattern in the past, summer storms must be considered the logical energy source for channel cutting in prehistoric time. It follows that the nature of summer rains in the Southwest commands the attention of geologists and other scientists who would use channel cutting and alluviation as a means of dating flood plain deposits. Considering its singular importance in characterizing the Southwestern environment, it is remarkable how little attention has been paid to the Southwestern (Mexican) monsoon.

II. SOUTHWESTERN CLIMATES AND THE MEXICAN MONSOON

The Southwest has a biseasonal rainfall. Summer rains (May-October) account for 60 to 70 percent of the annual precipitation in southeastern Arizona (127) and over 70 percent in eastern New Mexico (53). To the east and south, in west Texas and in the Mexican Plateau, this value rises; to the west and northwest summer rains diminish rapidly, accounting for 35-45 percent of the annual total in southeastern California (162, 168) and 44 percent at Salt Lake City (53).

The influx of Caribbean air masses into southern New Mexico and Arizona in July and August was recognized by Huntington (92) as a monsoon. Bryson and Lowry (30) summarized synoptic climatology of the Southwest describing the 500 mb. pressure changes.

In late June the high pressure cell over the eastern Pacific moves rapidly to the northeast and about July first the Southwest receives ". . . a deep, gentle flow from the Gulf of Mexico on the southwest side of the westward extension of the Bermuda High." Widespread rains begin and persist for several weeks. Moist tropical air masses from the Gulf of Mexico flow from Mexico across Arizona and New Mexico into Colorado and Utah. High intensity summer rainfall diminshes northward from the Mexican border (53).

Both the time of arrival and the expected amount of summer precipitation are less variable than the arrival and amount of winter storms (92). For Bowie and Tucson, two stations in southeastern Arizona, McDonald computed coefficients of variation of 0.62 ± 0.05, 0.54 ± 0.04 (winter precipitation) and 0.42 ± 0.04, 0.40 ± 0.3 (summer precipitation — a quantitative demonstration of Huntington's conclusion. Biologically speaking, this means that the summer annuals and their associated fauna should experience fewer seasons of drought-imposed stress than the annuals germinating after winter precipitation.

The relative amount of summer and winter rainfall has a profound effect on the Southwestern biota. Ensenada (32° N. lat.) on the Pacific Coast in Baja California and Monclova (27° N. lat.) on the Atlantic slope of northeastern Mexico lie at about the same latitude and receive about the same mean annual rainfall (340 mm. and 386 mm. respectively), Ensenada largely in the winter, Monclova in the summer. Shreve (168) stated that: "It is doubtful if a single native plant is common to the floras of Ensenada and Monclova. The differences in the character of the vegetation must be attributed largely to the monthly distribution of rainfall."

Southern Arizona falls under the influence of the two divergent precipitation systems. Following heavy winter rains the desert near Tucson and Phoenix is carpeted with flowers typical of California and the Great Basin — Cruciferae, Geraniaceae, Boraginaceae, Umbelliferae, Papaveraceae and Hydrophyllaceae. In late August, after summer rains, the desert flora is totally different, appropriately Mexican, and the preceding families are largely replaced by Chenopodiaceae, Amaranthaceae, Nyctaginaceae, Convolvulaceae, Martyniaceae, and Zygophyllaceae. Shreve (169) has described in detail the summer and winter annuals in various parts of the Sonoran Desert. He concludes that the opposing pattern of distribution, so clearly related to the seasonal distribution of rainfall, represents a geographic and climatic segregation that has endured for a long period of geological time.

In the desert grassland above 1,000 m. summer rainfall is critically important to the vegetation, determining forage production (141, 40) and the fate of the livestock industry. It was also critically important to the prehistoric spread of agricultural settlements (25, 10). In regions of deficient summer rainfall such as the California coast and central Nevada a corn-bean-squash economy was unknown prehistorically.

McDonald (pers. corr.) believes that the Southwestern monsoon is controlled by continental heating in the Midwest and Great Lakes region. An increase in summer temperatures in the Midwest would intensify the indraft of maritime tropical air masses from the Caribbean into Mexico and the Southwest, *increasing* summer precipitation. In an analysis of Mexican precipitation, Wallen (176) noted a rise in summer precipitation between the turn of the century and the mid-1930's. This ". . . must have been associated with a strengthening of the trade circulation over the Mexican area, this being, in its turn, caused by a general shifting polewards of the subtropical high-pressure cells and a transition to the west of the summer cell in the Caribbean Sea."

In a time series analysis of Arizona and western New Mexico Sellers (160) found a major maximum in July precipitation for 1907-1932, about the same time period which Wallen reported in Mexico. Sellers would also relate midsummer rains to thunderstorms and convection showers derived from moist tropical air and originating in the Gulf of Mexico and Atlantic Ocean. Sellers apparently agrees with Wallen that midsummer rains are most abundant when the semistationary high pressure cell centered in the Atlantic Ocean is displaced north and west of its normal position.

Sellers noted a relatively minor complication, the fact that heavy rains in September are associated with tropical disturbances in the Pacific Ocean. These ". . . are most likely to affect the Southwest when they are accompanied at high levels by a deep trough of low pressure extending southward from middle latitudes along the Pacific Coast" (160). At such times the Atlantic high pressure cell is weakened and displaced south and east of its normal position. Lag correlation coefficients show a relationship between September precipitation and that of the preceding winter, an expectation in accord with climatological theory (160).

With the qualification that September tropical storms may be responding to a different pressure condition, it appears that the normal summer convectional storms which have a profound influence on the biota and hydrology of the Southwest are dependent on the position of the Atlantic High. When the Atlantic High moves northward during periods of high latitude warming it is likely to bring a strong indraft of maritime tropical air into the Mexican Plateau and the Southwest. The summer is wet. When the Atlantic High is displaced southward, as happens during abnormally cold years in the north temperate latitudes, summer in the Southwest may be dry. Was this also the case in prehistoric time? What was the climate of the altithermal, 4,000 to 7,500 years ago?

The fossil record appears our main hope of pursuing the matter further. The abundance of pollen in flood plain and other postpluvial age deposits in arid regions provides us with a rich source of evidence. However, proper evaluation of fossil pollen requires a considerable familiarity with plant ecology. Before presenting the fossil pollen record of Southwestern flood plains, I will attempt to highlight some of the features of vegetation and the modern pollen rain which may be useful in interpreting the fossil record. The method is simple. Can we locate natural environments which yield a pollen rain comparable to that found in the prehistoric alluvium? If so, we can discuss more critically the nature of the prehistoric environment and climate.

III. VEGETATION OF THE SOUTHWEST

The pollen analyst is seldom able to identify plant species by their fossil pollen grains. Under ideal conditions, and by working with fresh material seen under phase contrast or the electron microscope, it is possible to achieve a high degree of accuracy in identification. But routine extraction and analysis of fossil pollen is far from an ideal condition for morphological work. I have found myself impatient to process and examine the common pollen types in many strata, as many as possible, rather than to undertake lengthy examination of pollen present in a few levels from a single locality. The kind of data rapidly and reliably gathered in routine pollen analysis is a collection of percentages of the relative abundance of easily recognized plant groups, notably the families or genera of wind-pollinated plants.

The task before the paleoecologist is to make a correct interpretation of the fossil count in terms of what is known about natural plant associations and their natural pollen rain. The procedure can be illustrated as an equation, where x represents the unknown fossil plant community:

$$\frac{x}{\text{fossil pollen count}} = \frac{\text{modern plant community}}{\text{modern pollen count}}$$

Problems in relating the pollen counts to plant communities involve an intimate knowledge of the local flora plus knowledge of density, abundance, tolerance, local distribution, and other ecological attributes of the plants found in the fossil record. There are many ways in which the fossil pollen count might be misinterpreted; badly needed are more data on how the present can be used as a key to the past, such as Potter and Rowley's (144) pioneering investigation of the relationship between vegetation and pollen rain in the San Augustin Plains of New Mexico.

To guide my interpretation of Southwestern plant communities I have drawn on some of the literature of Southwestern plant ecology, which begin to grow rapidly in the early 20th century under a series of contributions by staff and visitors at the former Desert Botanical Laboratory.[1] Not all key contributions came from botanists; the pioneering study of riparian communities was the work of a geologist (130) and the

classic analysis of biotic zones was proposed by a zoologist (134). What are the natural plant communities of the Southwest and how do they respond to environmental controls?

Biochores and Formation-Classes

Dansereau (44) would subdivide world vegetation into four very broad basic units or biochores: forest, savanna, grassland, and desert. Distinguishing desert, grassland, savanna, and forest in the fossil pollen record is the first problem of the paleoecologist. It is not always as simple as one might wish, especially when ecologists themselves are not in agreement about what constitutes a forest or a desert! The biochores may appear quite elementary, but their logical recognition on a world basis requires a determined disregard for the *species* present in the community under consideration, and close attention to the structure of the community, an outlook for which many biologists are unprepared. For example, under Dansereau's system of biochores pinyon-juniper woodland in New Mexico, coast live oak woodland in California, spruce-heath taiga in Labrador, and *Acacia* grassland in Africa are all grouped together as savannas within the savanna biochore. Despite the taxonomic differences, these climax communities share an obvious and distinctive structural attribute — scattered low trees and a continuous ground cover.

In arid regions with profound relief, such as southern Arizona (sea level to over 3,300 m.), examples of each of Dansereau's biochores may be seen within a few miles. Marshall's map (115) presents the distribution of coniferous forest, pine-oak woodland, encinal, grassland, and desert in southern Arizona and northern Sonora. Within Arizona, Nichol (142) maps nine major natural vegetation types which include at least six formation-classes. In northwestern Mexico pinyon-juniper and chaparral communities do not occur; arid tropical formations appear in southern Sonora. Otherwise, the vegetation zones of Arizona mapped by Nichol (142) can be traced into Mexico (17, 103) where they have their center of origin.

Forest Biochore (needle-leaved forest). Well-drained sites above 2,400 m. in the higher mountains of southern Arizona are dominated by coniferous forest. *Abies concolor, Pseudotsuga menziesii, Pinus strobiformis,* and

---

[1] Built on Tumamoc Hill at the edge of Tucson in 1903 by the Carnegie Foundation and presently occupied by the Geochronology Laboratories of the University of Arizona and the Rocky Mountain Range and Forest Experiment Station.

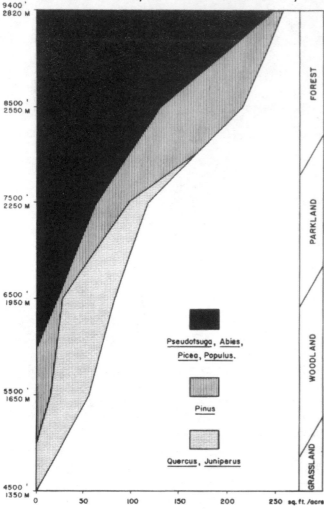

Figure 2. Basal area of trees on north slopes of Chiricahua Mountains.

P. ponderosa dominate most stands, commonly in mixtures of two to four of these species, with *Abies* and *Pseudotsuga* more important in the more mesic and the pines in the more xeric situations. Boreal forests of spruce and fir also occur, but are restricted to highest elevations of a few ranges. *Picea engelmanni* occurs in the Pinaleño and Chiricahua Mts., *Abies lasiocarpa* in the Pinaleño and Santa Catalina Mts. Except for local, successional stands of *Populus tremuloides*, broad-leaved deciduous trees are not dominant in the mountain forests. In addition to *Populus* the deciduous species often present in coniferous forests are *Acer glabrum*, *A. grandidentatum*, *Quercus gambelii*, *Salix scouleriana*, and *Robinia neomexicana*.

At lower elevations, from 1,800 to 2,400 m., yellow pine *(Pinus ponderosa* or *P. arizonica)* predominates. Major associates are *Pinus strobiformis* and *Pseudotsuga*, which increase in importance toward higher elevations and toward more mesic sites, while three oaks *(Quercus reticulata, Q. hypoleucoides, Q. arizonica)* increase in importance toward lower elevations and more xeric sites. The yellow pine "forests" are decidedly more open, admitting much more sunlight to the forest floor, than the mixed coniferous for-

ests above them. Before the present program of fire control was initiated about 50 years ago, the tall open stands of *Pinus ponderosa* which extend through the Southwest were even more open, with a rich bunch-grass cover (39). In their primeval condition the yellow pine stands are better classed as a tall tree parkland or savanna rather than as a closed canopy forest.

Within the mountain forests of the Chiricahuas and Pinaleños are occasional natural meadows. Here *Iris missouriensis, Helenium hoopesii,* and *Artemisia dracunculoides* overlie fine soil one to two meters in depth. Weathered rock lies below. Although the meadow soil looked promising for pollen analysis, the extraction of samples collected from the meadow at Rustler's Park, 2,580 m., in the Chiricahuas yielded little beyond a great quantity of finely divided charcoal. A few conifer pollen grains were present but insufficient for a pollen count.

Savanna Biochore (pine-oak woodland and encinal). In an attempt to describe tree abundance from mountain-top forest to desert grassland, I measured tree trunk density on north slope communities through the middle of the Chiricahua Mountains. Results of plotless sampling at each 300 m. are summarized in Fig. 2. Basal area of tree trunks increases from zero at 1,300 m. to 250 sq. ft. per acre at 2,800 m. Coniferous forest and tall tree parkland with more than 75 sq. ft. basal area per acre prevail above 2,000 m. Below this elevation the conifers decrease in both basal area coverage and in relative importance in the tree stratum. The oaks, which are present in the ponderosa pine forests, increase in relative importance, so that the "savanna biochore" is characterized here by dominance of evergreen-sclerophyll oaks.

In the upper part of the biochore, the oaks *(Quercus hypoleucoides, Q. arizonica, Q. emoryi)* are of moderately high basal area and crown coverage and share dominance with pines — usually scattered individuals rising above the oak canopy. The pines may include *Pinus leiophylla, P. cembroides, P. engelmannii* and in Mexico *P. durangensis. Arbutus arizonica, Juniperus deppeana, Arctostaphylos pungens, A. pringlei,* and *Yucca schottii* are other associates. This is the pine-oak woodland of Marshall (115).

Toward lower elevations and on drier sites basal area and crown coverage decrease until stands are very open, with round-crowned oaks *(Quercus arizonica, Q. emoryi)* and at lower elevations *Q. oblongifolia* and *Vauquelinia californica* scattered above an undergrowth of grass or shrubs *(Arctostaphylos, Yucca, Garrya wrightii, Nolina microcarpa, Dasylirion wheeleri, Agave palmeri, Haplopappus laricifolius).* This is the encinal of Shreve (162), a small-tree savanna, though in some areas its undergrowth is dominated by shrubs rather than grasses.

The pine-oak woodland gains in species and extends its altitudinal amplitude as one proceeds from central Arizona to central Sonora; its character is elegantly described by Marshall (115). In structure, parts of the pine-oak woodland resemble the Mediterranean savanna of Dansereau (44). It must be emphasized that evergreen oak woodland is well developed in certain regions of summer rain and winter drought (the Mexican Plateau), as well as in Mediterranean climatic areas with a summer drought and winter rain (Portugal, California). The same is true of chaparral, so often but incorrectly considered exclusive to a Mediterranean-type climate (138).

Grassland Biochore (desert grassland or mesquite grassland). Perennial grasses and forbs with trees confined to washes occupy bajadas of southern Arizona and southern New Mexico between 1,200 and 1,500 m. West of the Chiricahuas, north of the Huachucas, and between the Santa Ritas and Whetstones in the Empire Valley, one finds extensive grass-covered plains. Formerly, some of the grasses (*Bouteloua rothrockii*) were cut for hay (172). The desert grassland of the Southwest is an extension of the Mexican mesquite grassland (17, 103). Scattered shrubs usually occur, including, in addition to the ubiquitous mesquite, species of *Acacia, Yucca, Agave, Haplopappus, Gutierrezia, Baccharis,* and *Ephedra.*

The dominant genera of grasses are *Bouteloua, Hilaria,* and *Sporobolus.* The perennials include *B. eriopoda, B. rothrockii, B. hirsuta, B. gracilis, B. curtipendula, B. filiformis,* and *B. chondrosioides, H. belangeri, H. mutica, S. airoides* and *S. wrightii.* In Cochise County Darrow (45) assigns about 200,000 acres to both mixed gramas and tobosa. In the Jornada Experimental Range of southern New Mexico the dominant grass of the plain portion is *Scleropogon* with *Bouteloua eriopoda, Hilaria mutica,* and *Sporobolus* very common (108).

Of 134 grasses listed by Gentry (67) near Durango City, 1,000 km. south of the Arizona border, 66 percent also occur in Arizona. The ten most abundant grasses in Gentry's quadrats are also dominants or common species in southern Arizona. Even further to the east in Mexico the grassland of Coahuila near Saltillo is dominated by familiar Arizona species, *Bouteloua curtipendula, B, gracilis, B. hirsuta, B. filiformis, Aristida,* and *Hilaria belangeri* (173). Rich development and diversity of the desert grassland south of the border bespeaks a Mexican origin.

In the desert grassland maximum productivity occurs in late summer following July-August rains. In southern New Mexico about 90 percent of the annual forage production occurs at this season (141), while on the Santa Rita Experimental Range of southern Arizona Culley (40) reported 83 percent of the summer's production of grass occurred between July 20 and September 10.

As a distinct vegetation zone the grassland biochore is confined to relatively level or rolling terrain.

On the south side of the Catalina Mountains the Sonoran Desert extends up to the lower edge of the encinal or open oak woodland at about 1,200 m. without an intervening grassland belt (162). On the precipitous western face of the Sierra Madre Occidental grassland fails to occupy its appropriate zonal position (166). Shallow soils within the grassland of southern Arizona are commonly occupied by *Prosopis, Acacia,* or *Juniperus.*

The transformation of the desert grassland into brushy savannas is perhaps the most dramatic alteration of an uncut, uncultivated North American plant community witnessed within the present century. An extensive increase in the frequency of shrubs, especially of mesquite (71) but also including burroweed, cane cholla, creosote, whitethorn, and others, has blighted cattle ranges in Arizona (143, 70, 89, 90) and New Mexico (65, 18). The "brush invasion" is commonly attributed to grazing effects; it may also reflect a reduction in fire frequency. According to Humphrey (89), "It seems probable that had fires not periodically swept the desert grassland, most, or perhaps all, of the area would have supported a woody overstory long before the first white man set foot on North America."

The brush invasion has added to the usual problems expected in vegetation analysis and is partly responsible for considerable discrepancies between various vegetation maps of the Southwest. In Arizona Marshall (115) interpreted grassland much more conservatively than Nichol (142); in southern New Mexico Shreve (167) and Benson and Darrow (11) mapped the Chihuahuan Desert through a region Castetter (33) shows as desert grassland. In extreme northwestern Chihuahua Shreve (165) extended "desert" to the Sonora border in a region Hernandez and Gonzalez (82) mapped as grama grassland. In northwestern Sonora Shreve (169) extends the foothills division of the Sonoran Desert into part of the region White (179) considered mesquite-grassland and far beyond the limit of creosote-palo verde-cacti desert of Brand (17).

In speaking of the Sulphur Spring Valley as "desert grassland," within the grassland biochore, one must recognize that roughly half the bajada slopes are dominated by *Prosopis, Larrea, Acacia, Flourensia,* or other shrubs.

Desert Biochore (Sonoran and Chihuahuan Desert). Southeastern Arizona lies between two great desert regions of North America, the low-lying Sonoran Desert of southwestern Arizona, Baja California, and Sonora (169) and the more elevated Chihuahuan Desert of the Mexican Plateau. In southeastern Arizona the two are separated for the greater part by desert grassland covering the bajadas of Cochise and Santa Cruz Counties.

Most of the present study area (Fig. 1) lies east of the region dominated by giant cacti, palo verde, shrubby *Franseria, Encelia,* and *Simmondsia,* thus east of the area mapped as Sonoran Desert by Shreve

(169). If Benson and Darrow (11) and Lowe (109) are correct in extending the Sonoran Desert to the New Mexican border in the San Simon Valley, the pollen profile of the San Simon Cienega could be considered representative of the Sonoran Desert.

The shrub-dominated communities of the Chihuahua-Sonora-New Mexico-Arizona boundary bear a much closer relationship to communities of the Chihuahuan Desert in central Mexico than to any part of the Sonoran Desert to the west. In the Chihuahuan Desert of Coahuila Muller (138) found "the most characteristic species is *Larrea tridentata* (DC.) Cov., and the structure and composition of the variant types may best be considered from the standpoint of the species that are associated with *Larrea* or that occasionally replace it." Widely associated with *Larrea* in both Coahuila and southeastern Arizona are *Flourensia cernua, Acacia vernicosa, Fouquieria splendens,* and *Prosopis velutina.* Of the following 17 less common species associated with *Larrea* in Coahuila, 11 occur in southeastern Arizona (marked with asterisk):

*Condalia lycioides	Yucca australis
*Koeberlinia spinosa	Yucca torreyi
Coldenia greggii	*Acacia constricta
*Parthenium incanum	*Rhus microphylla
*Lycium berlandieri	Citharexylum brachyanthum
*Celtis pallida	*Microrhamnus ericoides
*Condalia spathulata	Sericodes greggii
Opuntia imbricata	*Hilaria mutica
*Opuntia leptocaulis	

Considering the distance (over 1,000 km.) between southeastern Arizona and southern Coahuila, eastern Mexico, the floristic homogeneity is notable.

Environmental Factors Affecting the Distribution of Upland Vegetation

The following outline indicates the major environmental factors known to influence the pattern of vegetation zones in the arid southwest.

Elevation. Merriam's famous report on the San Francisco Peaks (134) illustrates elegantly the profound effect of elevation on plant and animal life. While Merriam's life zone abstraction exerted a profound effect on ecologists of his generation, the uncritical correlations required by the life zone concept provoked a wave of criticism (47). Merriam's temperature laws were generally rejected. However, all ecologists recognize that profound changes in the vertical distribution of natural vegetation and of most species of animals and plants invariably accompany changes in elevation. Merriam's descriptions of zones in the San Francisco Peaks area are quite good and have not been improved. In southern Arizona the arrangement of vegetation is slightly different. The Sonoran Desert extends from sea level to 3,500 feet (1,050 m.), the desert grassland to 5,000 feet (1,500 m.), pine-oak woodland to 7,000 feet (2,100 m.), pine parkland and forest to 10,000 feet (3,000 m.), and spruce forest occupies certain mountains above 9,000 feet (2,700 m.).

In his climatic and botanical study of the Santa Catalina Mountains Shreve described profound vertical changes roughly comparable to those found by Merriam on the San Francisco Peaks. Shreve abstracted them into a series of biochores (forest, woodland or encinal, and desert), but did not fail to stress the individualistic nature of plant distribution. "It is nowhere possible to pick out a group of plants which may be thought of as associates without being able to find other localities in which the association has been dissolved" (162). In this respect Shreve anticipated the continuum concept of modern ecology.

Latitude. It is widely recognized that as one proceeds south along the Rocky Mountains various plant communities, including the lower limit of trees, ascend to successively higher levels. On the other hand, within Arizona and New Mexico the lower limit of most tree growth *does not continue to rise* as one proceeds southward. Instead, it drops. This apparent anomaly has been known at least since the time of Rothrock (148). In Arizona from the Mogollon Mesa in Coconino County to Camp Crittenden in Santa Cruz County, a distance of about 300 km., Rothrock reported a decline in the lower limit of pine of about 1,000 feet (300 m.). In southern New Mexico yellow pine communities also occur 1,000 feet below their lower altitudinal limit in the northern part of the state, a feature Antevs (4) realized would complicate paleoclimatic interpretations. Pine, oak, and fir continue to descend in elevation as one proceeds southward from Arizona (115). Pine and oak approach sea level in southern Mexico; fir reaches 1,300 m. in cloud forest of eastern Mexico (116), 800 m. *below* its lower limit in southern Arizona. While the rise in lower limits of montane plants from the northern to the southern Rockies is a function of increasing mean annual temperatures, the fall in elevation at lower latitudes from Arizona and New Mexico into Mexico may be related to increasing precipitation.

For the palynologist this means that it is not immediately obvious what type of climatic change (whether warm-wet, cold-wet, or cold-dry) is expressed by a past increase in pollen of pine and other montane mesophytes. *In the Southwest any major increase in moisture will depress all montane vegetation zones* — the position of Southwestern vegetation zones represents an interaction of precipitation and temperature.

Slope Exposure. The ecologist learns to expect basic biotic change in response to change in elevation and latitude. Only slightly less important in the arid Southwest is the effect of slope exposure. Merriam (134) found that the normal average difference in altitude of a given zone on the southwest and northeast sides of the San Francisco Mountains is about 275 m. In the Catalina Mountains the north slope-south slope difference is from 180 to 300 m. at lower elevations and from 300 to 600 m. at upper elevations (162). In Utah Woodbury (184) diagrams maximum differences in north-south slope communities of over 700 m.

In the Chiricahua and Pinaleño Mountains of southern Arizona the mesic species of high elevations, *Picea engelmanni* and *Abies lasiocarpa,* occur only on sheltered northerly slopes and ravines above 2,550 m. and are nowhere seen on the south side. On Green's Peak of the White Mountains in eastern Arizona spruce also occupies the north slope, while the south slope is completely treeless. In the Subalpine (spruce) zone of Utah Ellison (57) observed: "The most extensive and densest spruce-fir stands are found on north exposures, and only small, scattered patches occur on level ground and south exposures" (p. 173). In montane forest of the Colorado Front Range, Marr (113) found that open stands of ponderosa pine on south slopes oppose closed forest on north slopes.

At low elevations in the desert grassland and in the Sonoran Desert itself the effects of slope are less conspicuous. But even here they can be found. Near Tombstone Keppel, et al. (97) report a significant difference in shrub cover in two of 12 species studied. *Larrea* was more abundant on the southwest slope and *Bouteloua curtipendula* on the north slope. At low elevations in the Chiricahuas Blumer (13) noted *Lippia wrightii* on south slopes near the upper limit of its range (1,800 m.) and just as definitely limited to north slopes near its lower limit (900 m. and lower). Near Tucson saguaro *(Carnegiea gigantea)* is much more abundant on south than on north slopes. Major north-south slope effects are found in arid regions of the Old World (16). They are not characteristic of the tropics and the Arctic, but within semiarid regions of mid-latitudes only the Alpine tundra zone fails to reveal north-south slope contrasts (113). The contrasts in vegetation one finds on opposing slopes may alter topography. In the intermontane region in Utah south-facing slopes tend to be long and gentle, north-facing slopes to be short and steep (56).

Mountain Sides. Major differences in vegetation unrelated to slope exposure may be encountered on different sides of the same range. For example, on the north side of the Mogollon Rim ponderosa pine and pinyon are replaced by more xerophytic communities far above elevations they occupy immediately south of the Rim. Shreve (165) reported a similar arrangement along the Sierra Madre Occidental where the sharply dissected west slopes draining into the Pacific (base elevation near sea level) have pines growing 350 to 800 m. below the lower limit of pine on the east side of the range bordering the central Plateau (base elevation 1,350 m.). Less spectacular mountain side differences are seen in the Rincon Mountains (13,114) and Huachuca Mountains (177). In these cases mesoyhptic communities occur at lower levels on the east side.

Shreve (164) concluded that differences in vertical distribution of vegetation on different sides of the same mountain were caused by base level differences. Merriam (134) considered base elevation more important than slope exposure and related the upward displacement of more xeric plant communities on the side of the range with the higher base level to a more intense diurnal heating. On the other hand, in the case of the Mogollon Rim and the Sierra Madre Occidental a rain shadow phenomenon is also likely to underlie the lack of symmetry.

The Merriam Effect. In his classic study of life zones on the San Francisco Peaks Merriam (134) illustrated the differences one finds in vertical distribution of trees growing on adjacent peaks of different mass. Merriam found spruce on the San Francisco Peaks growing at an elevation 150 m. below the top of adjacent, smaller O'Leary Peak where spruce is absent. Shreve (164) noted that oak-pine woodland is missing from Black Mountain, despite the fact that the top of Black Mountain coincides with the vertical distribution of oak-pine woodland in the Catalinas, a larger adjacent range.

In southern Arizona spruce descends below 2,700 m. on north slopes of the relatively bulky Chiricahuas and Pinaleños. Although they exceed 2,700 m. in elevation the more attenuate, less bulky Huachucas, Santa Ritas, and Santa Catalinas harbor no spruce trees (110). In recognition of Merriam's discovery Lowe (110) has designated the relationship between mountain mass and the vertical distribution of animals and plants the "Merriam effect."

The Merriam effect can be seen in the vertical distributions of all trees occupying the desert mountains. The Chiricahuas and Huachucas are adjacent isolated ranges in southern Arizona; the Chiricahuas are only 106 m. higher but are considerably more massive. Data on vertical distribution of common trees were gathered in the Chiricahuas in the summer of 1956. The results were compared with vertical distribution data of trees in the Huachucas made by Wallmo (177). There are differences in both the lower and the upper limits of trees occupying both mountains (Fig. 3); the difference in lower limits increases with increasing elevation (Fig. 4). The seven species of trees occupying relatively low elevations in both ranges reach virtually the same lower altitudinal limit; those in the Chiricahuas occur only slightly lower, 30 to 150 m. below those in the Huachucas. For the seven species of trees found at higher elevations the difference between mountains is much greater; in the Chiricahuas these trees occur 300 to 500 m. below their lower limit in the Huachucas (Fig. 4).

To the pollen analyst the Merriam effect means that the mass of mountain ranges near a pollen sampling point will make a greater difference in the amount of exotic pollen (such as pine or spruce) reaching a mid-valley deposit than one might expect in terms of life zone theory alone. If adjacent mountains differ in bulk

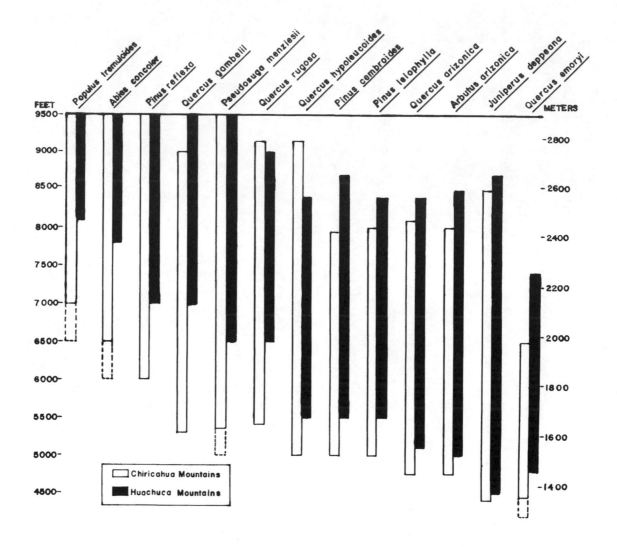

Figure 3. Vertical distribution of trees in Chiricahua and Huachuca Mountains.
Dotted lines indicate rare occurrences.

the same elevation will not be occupied by the same plant community.

The Merriam effect is not unique to arid regions. Van Steenis (170) writes of it in Malaysia: "The larger the mountain massives the more broad in a vertical sense are the physiognomical and altitudinal zones." Van Steenis found that stenotherm temperate mountain plants, those rarely descending below 1,000 meters, are often missing on peaks which attain only 1,500 or even 2,000 m. elevation.

In the Great Smoky Mountains of eastern United States spruce-fir forests which occupy slopes above 1,400 m. in the northeastern half of the range do not appear on smaller peaks rising to around 1,700 m. in the southwestern half of the range. Whittaker (180) attributed the absence of spruce and fir on the small peaks to postglacial climatic change. He suggested that warmer climates during the hypsithermal displaced the spruce-fir zone upward by 400 m., exterminating spruce and fir in the area of lower peaks while they persisted

on the sanctuary offered by the higher peaks of the northeast half of the range. The Merriam effect was not thought responsible for the contrast because vegetation patterns at 1,200-1,400 m. were closely similar in the two areas, suggesting similar local climates, and because the north slopes and draws in the southwest half of the Smokies should offer environments for spruce more mesic than the south slopes occupied by spruce in the northeast half, even if the general climate of the southeast half above 1,400 m. were somewhat drier.

Thus, it would appear that in the Great Smoky Mountains the more probable explanation for absence of expected spruce-fir communities on certain peaks is a postglacial climatic change — an altithermal displacement of biotic zones — rather than local climatic differences related to mountain mass.

In the isolated Sierra de Tamaulipas of eastern Mexico a xerothermic (altithermal) effect was also postulated to explain the lack of typical pine-oak woodland mammals and lizards in a habitat considered to be

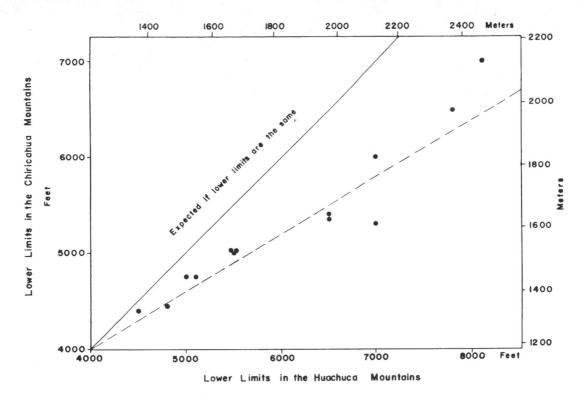

Figure 4. The Merriam effect in the Chiricahua and Huachuca Mountains. Downward displacement occurs in the more massive range, the Chiricahuas.

suitable for them (120). In Africa the same explanation was offered to account for the absence of an Alpine zone in the mountains of Biafra (15).

In the Southwest the hypothesis of upward displacement and elimination of upper zone species during the altithermal will not explain the nature of present montane vegetation. Noting that relict populations of pinyon occupy the *top* of isolated desert ranges in the Great Basin, Aschmann (10) concluded that climatic conditions in postpluvial time could not have been appreciably hotter and drier than they are at present. Aschmann's observation applies equally to southern Arizona where one can find a variety of very small populations of trees and shrubs clinging to sheltered north-facing microhabitats at the crest of various ranges — *Quercus turbinella* at the top of the Tucson Mountains, *Pinus cembroides* at the top of the Mule Mountains, and *Abies lasiocarpa* on the north side of the top of the Santa Catalina Mountains.

The mountain top tree populations give the appearance of barely surviving under present climatic conditions. Unless they arrived relatively recently (in the last few thousand years) they may be viewed as evidence that a major upward displacement of biotic zones has not occurred during the postglacial period. Differences between zones on adjacent peaks is satisfactorily attributed to local differences in mass and climate — the Merriam effect.

Edaphic Conditions. Bedrock composition has long been recognized as controlling vegetation. In the Southwest Shreve (164) reported: "The lowest absolute elevations are reached by encinal and forest on gneiss and granite, and desert forms correspondingly attain lower maximum elevations on those rocks and their derived soils." Forest and encinal are displaced upward on basalt, rhyolite, and other volcanics and retreat still farther on limestone. In the San Francisco Peaks area of northern Arizona, a region of volcanic rock, pine and juniper grow on the relatively fresh ash from Sunset Crater at a level about 300 meters below their lower limit on other soils (37). Sand deposited on the shores of playa lakes in the Southwest may favor tree growth below the normal limit of certain species, for example *Juniperus deppeana* and *Quercus* in grassland on the northeastern shore of Cloverdale Playa and *Pinus ponderosa* on the northeastern edge of the San Augustin Plains.

Important differences between upper and lower bajada vegetation near Tucson are readily explained in terms of differences in soil texture (186). Edaphic differences between separate bajadas of the same valley may be expected when their mountain sources differ in bedrock composition. For example, *Hymenoclea* and *Ephedra,* shrubs partial to sand, are much more abundant on the sandy bajadas derived from gneiss of the Catalina Mountains than on the stony bajadas derived from rhyolite of the Tucson Mountains. Near the Chiricahua Mountains *Larrea* occurs commonly only on bajada soils derived from limestone.

Inversions. Deep, narrow ravines produce temperature inversions and ground water conditions favorable to montane mesophytes. In Bear Canyon of the Catalinas and in Rucker Canyon of the Chiricahuas, yellow pine forest grows below encinal. Populations of fir and spruce

11

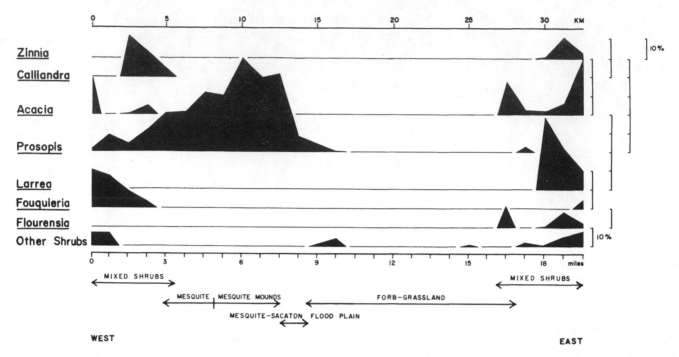

Figure 5. Shrub coverage in percent cover across Mule and Swisshelm bajadas.

in Oak Creek Canyon occur far below the lower limit of these trees on the open slopes of the San Francisco Peaks 30 km. to the north. *Quercus emoryi, Q. oblongifolia,* and *Q. arizonica* descend to 820 meters along the flood plain of Sabino Canyon, 550 meters below their lowest occurrence on slopes in the Catalina Mountains (162). These deviations follow the rule discussed by Daubenmire (48) that each zone attains its lower altitudinal limit in valleys and its upper limit on ridges, bringing about zonal interfingering. The effect is prob-

Figure 6. Diagrammatic transect of vegetation and valley fill across Mule and Swisshelm bajadas of the Sulphur Spri

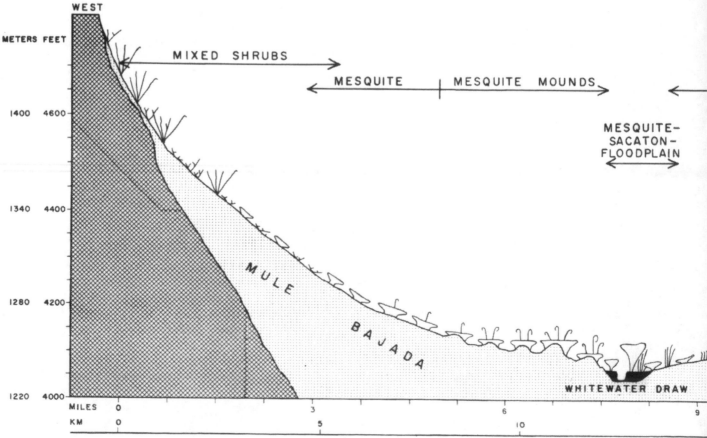

ably greatest in arid subtropical latitudes.

Fire. Fire may alter the character of grassland, woodland and forest but it is unable to spread through the desert. In yellow pine parkland frequent light burns increase both the perennial grass cover and the rate of pine growth (39). Lacking fire protection, the montane pine parkland of northern Mexico is more open than homologous stands in the United States (115). The Mexican pines occupy a deep soil with a rich cover of bunch grass. In the desert grassland the reduction in range fires probably accounts in part for the spectacular brush invasion of the last 50 years (89).

To summarize, I have commented very briefly on the Southwestern environment. It appears that some factors, i.e. slope exposure, mountain mass, bedrock composition, and climatic inversions, may have more effect on the distribution of plants in arid than in humid regions. All of them can alter the character of the natural pollen rain and therefore all command the attention of the pollen stratigrapher.

The vegetation pattern responds to many variables, but if one confines observations to a single slope of uniform substrate, a consistent pattern of change will emerge. In Arizona and New Mexico, on those mountains where they occur together, one invariably finds *Quercus emoryi* growing slightly below *Q. arizonica, Q. hypoleucoides* below *Q. gambelii, Pinus ponderosa* below *P. reflexa, Pseudotsuga* below *Abies,* and *Picea pungens* below *P. engelmanni.* Such experience led Blumer (14) to claim: "In the mountains, without instrument or map, one can determine the altitude within a very few hundred feet by the species he meets on the slopes." Can the palynologist approach this level of precision in his fossil studies?

Pollen Sources in the Desert Grassland

Having proposed a framework for classifying the major vegetation units in the Southwest and having discussed the behavior of vegetation zones under varying ecological conditions, I will now focus attention on the vegetation zone within which most of the study areas are located — the desert grassland. To obtain some quantitative information regarding coverage and community composition, and in hope of gaining insight into upland as opposed to flood plain sources of pollen in the modern pollen rain, data was collected on shrubs and other plants along a transect across the Sulphur Spring Valley. Estimates of shrub crown coverage can be obtained quickly by the variable plot method described by Cooper (38). Except for an area of sandy mounds covered by thickets of mesquite in which individual shrubs could not be recognized (Photograph 7), Cooper's method proved rapid and effective.

The west end of the transect lies at 1,300 m., 0.4 km. northwest of Dixie Canyon Ranch, T. 22 S., R.

alley. Inner valley alluvium shown in black; bedrock is cross-hatched; valley fill is stippled. Vertical exaggeration 40 times.

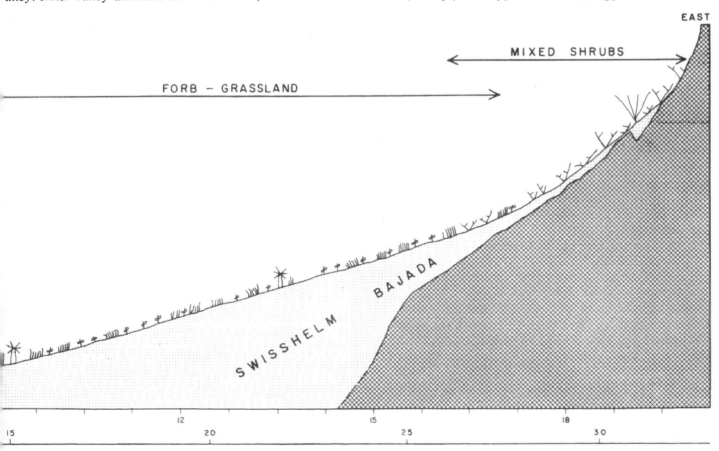

24 E., northeast corner of sec. 24, at the foot of the Mule Mountains. The transect runs roughly parallel to the slope of the Mule and Swisshelm bajadas. It crosses the Whitewater Draw about 10 km. north-northwest of Double Adobe. The eastern end is in T. 21 S., R. 28 E., sec. 19.

The transect revealed a complete lack of equivalence between communities on the eastern (Swisshelm) and western (Mule) bajadas (Figs. 5 and 6). The former features a very gentle grade, fine soil, and relatively poor dissection. Present-day agriculture (all by irrigation with ground water pumping) is concentrated on the east side of the valley, close to the flood plain of the Whitewater Draw. Desert grassland dominated by *Hilaria mutica, Aristida, Croton, Bahia,* and other annual forbs occupies the uncultivated parts of the Swisshelm bajada (Photograph 10). At comparable elevations on the Mule bajada south of the El Paso Gas Company pipeline there is no forb-grass dominance. Strings of mesquite, *Prosopis juliflora torreyana,* about a meter in height and up to 20 m. in length, sprawl across the lower Mule bajada (Photograph 7). They are surrounded by a bare hardpan which resists invasion of all plants, even the summer annuals. Such annuals as do occur abundantly *(Chloris, Amaranthus palmeri, Evolvulus)* grow on the mounds where they must compete with the mesquite.

Ninety species of herbs were collected along the transect shown in Fig. 6. As in the case of the shrubs, there are many species of herbs which are exclusive to or much more abundant on one side of the Sulphur Spring Valley than on the other. Found on both the Mule Mountains bajada and the Swisshelm Mountains bajada are the following: *Allionia incarnata, Amaranthus palmeri, Apodanthera undulata, Aristida adscensionis, A. longiseta, Bahia absinthifolia, Baileya multiradiata, Boerhaavia coulteri, B. erecta, Bouteloua aristidoides, B. rothrockii, Calliandra humilis, Chloris virgata, Croton corymbulosus, Cucurbita foetidissima, Drymaria sperguloides, Euphorbia capitellata, E. serpyllifolia, Jatropha macrorhiza, Kallstroemia grandiflora, Muhlenbergia porteri, Perezia wrightii, Proboscidea parviflora, Solanum rostratum, Tidestromia lanuginosa, Tridens pulchellus, Verbesina encelioides,* and *Zinnia grandiflora.*

Species found only on the Mule Mountains bajada include: *Alternanthera repens, Aster tanacetifolius, Boerhaavia intermedia, Bouteloua chondrosioides, Eragrostis arida, E. megastycha, Eriochloa lemmoni, Euphorbia albomarginata, E. serrula, Froelichia gracilis, Hibiscus denudatus, Hymenothrix wrightii, Ipomoea hirsutula, Lepidium thurberi, Menodora scabra, Mollugo verticillata, Pectis filipes, P. prostrata, Physalis wrightii, Portulaca umbraticola, Senecio* sp., *Setaria macrostachya, Sida procumbens, Trichostema arizonicum,* and *Tridens muticus.*

Species found only on the Swisshelm bajada are the following: *Acalypha neomexicana, Amaranthus graecizans, Aristida divaricata, A. glabrata, A. turnipes, Berlandiera lyrata, Boerhaavia wrightii, Bouteloua barbata, B. eriopoda, Cassia bauhinioides, Commelina erecta, Conyza coulteri, Crotalaria pumila, Epilobium* sp., *Eriogonum albertianum, Eupatorium greggii, Euphorbia revoluta, E. stictospora, Franseria acanthicarpa, Gaillardia pulchella, Gaura parviflora, Gilia longiflora, Hilaria mutica, Hoffmanseggia densiflora, Kallstroemia californica, Lepidium* sp., *Mentzelia pumila, Panicum obtusum, Sanvitalia aberti, Scleropogon brevifolius, Sida leprosa, Sporobolus airoides, Talinum auranthiacum, Trichachne californica, Trianthema portulacastrum, Xanthium saccharatum, Zinnia pumila.*

Shrub communities of the upper bajadas resemble each other more closely than do communities of the opposing lower bajadas (Fig. 6). There is a marked increase in both density and number of shrub species as one approaches the mountain backslope. Nevertheless, even between the two upper bajadas there are community differences, probably related to soil differences. Near the Mule Mountains ocotillo, mesquite, *Calliandra* and *Acacia* prevail (see Fig. 5 and Photograph 6). The top of the Swisshelm bajada is dominated by *Larrea, Acacia,* and *Flourensia. Gutierrezia* and *Haplopappus* occupied transect stations on the Mule side only; *Zinnia, Parthenium,* and *Menodora* were only on the Swisshelm side.

Among the shrubs of the Sulphur Spring Valley only pollen of *Ephedra* and *Prosopis* is regularly encountered in samples of stock tank mud or fossil alluvium. Despite their local abundance (see Fig. 5), I have found only one pollen polyad of *Calliandra,* one pollen grain of Cactaceae (flood of Dixie Wash), and two of *Acacia. Zinnia, Haplopappus, Gutierrezia, Flourensia,* and *Parthenium* might have contributed to the composite pollen curve; it is not possible to separate them from other composites which are annuals. In brief, the prospect for following regional history of the dominant upper bajada shrubs is discouraging.

Regarding the herbs, in the Double Adobe area grasses dominate on the Swisshelm bajada, while *Amaranthus palmeri,* an abundant pollen producer, is most numerous on the Mule bajada, where its scarious spikes project above the mesquite mounds (Photograph 7). The wind-pollinated Compositae are nowhere abundant. In small bare clay depressions *Xanthium* springs up after summer storms (Photograph 8). Annual *Franseria* occupies mud flats of the Whitewater Draw.

The common pollen producing herbs of the desert grassland (the families Compositae, Gramineae, and Chenopodiaceae plus *Amaranthus*) include many species which grow in both the flood plains and on the bajadas. Whether caused by arroyo floods of the summer rains, or by trampling and water leakage around cattle tanks, disturbance of soil definitely favors the chenopod-amaranth group. This may account for its dominance in most cattle tank and alluvial samples at present. On undisturbed soils of the upper bajadas

Compositae pollen may dominate. From soil surface pollen studies it would appear that most of the non-arboreal pollen preserved in alluvial sediments was produced by vegetation growing very close to the sampling point. Other than small amounts of pine, oak, and *Ephedra* the pollen rain is not known to include either upper bajada or montane sources. In brief, today the major source of the pollen found in alluvial deposits of the desert grassland is from plants growing on the flood plain itself.

IV. THE MODERN POLLEN RAIN

What is the present pollen rain of the Southwest; how are we to relate it to the past? Can we determine the source of pollen trapped in flood plain sediments? What is its ecological significance?

Modern pollen samples reveal the relationship between vegetation and the local pollen rain. A ready source is the ooze (sludge or mud) contained in the bottom of metal-rim stock tanks (Photographs 3 and 4). In addition, a rich pollen record can be found in the soil surface. Pollen is present in rain, runoff, fresh arroyo alluvial deposits, and dung. Results of modern pollen counts are presented in Table 1 and in Fig. 7.

Stock tanks are maintained by most ranchers throughout the desert grassland. Those below 900 m. in elevation are often covered to prevent evaporation; they do not serve as pollen traps. In the desert grassland there are many open-top, metal-rim tanks about two meters high (above the reach of livestock) and five meters in diameter, permanently filled and trapping atmospheric particles at all seasons. Unless fossil pollen is carried into the tank by the ground water, the only pollen source is wind transport, either primary pollen fallout or secondary fallout following wind deflation of dry pollen on the soil surface.

Both stock tanks and flood plain alluvium reflect the pollen rain of relatively disturbed sites rather than the "natural" rain of undisturbed upland communities. Disturbance around tanks is caused by intense trampling and a generous spread of manure from livestock. Following summer rains the stock tanks are enclosed by rank patches of *Amaranthus, Salsola,* and other weedy annuals (Photographs 3 and 4) also typical of flood plains.

Another popular type of stock pond is the "dirt tank," an earth dam across a gully which traps runoff from storms. Between rains the accumulated sediment, rich in pollen, dries out and cracks. Sediment rapidly fills the dirt tanks, and presumably much of the pollen they contain was washed in from adjacent upland soil surfaces.

A major problem regarding history of the flood plains is the source of pollen they contain. In lakes or bogs within forest regions one assumes that most of it was wind-borne. In the Orinoco Delta Muller (139) has shown the importance of water transport, and the question arises concerning the role of runoff in Southwestern pollen sedimentation. Analysis of sediment in flood water following August storms leaves no doubt that pollen is carried by this means (Table 1).

Fortunately, the method of transport may be unimportant, at least in gross interpretation of fossil results. At the top of Fig. 17 are pollen counts of the ooze of metal-rim tanks which resemble closely pollen counts of adjacent flood plain alluvium deposited within historic time. Both are dominated by cheno-ams.

One of the assumptions underlying pollen stratigraphy of temperate regions is that regional mixing will obliterate local community segregates. For example, a dense stand of hemlock *(Tsuga)* on a north slope above a lake will not be greatly overrepresented in the pollen content of adjacent lake sediments. In flood plain alluvium of the Southwest regional mixing does not occur. Beneath or close to colonies of giant ragweed in the Empire Cienega the frequency of Compositae pollen is 70 to 80 percent. 120 meters from the giant ragweed the frequency of Compositae drops to less than three percent and cheno-ams account for 90 percent of the relative pollen count (Table 3, Fig. 13).

Regional mixing is more likely to characterize tree pollen wafted from montane forests into the desert grassland. When local pollen production is extremely poor as in the creosote bush desert of the Panamint Valley, and when there is an abundant source of pine pollen upwind in adjacent mountains, the soil surface of the desert may contain 40 to 50 percent pine pollen.

Pine Frequency and Pine Size Frequency. Utilizing the analysis of the stock tanks, it is possible to relate pine pollen frequency to elevation. The pine pollen record is crucial to the interpretation of past climatic changes in the Southwest. As would be expected, pine frequency is dependent on elevation and on distance from source trees. The pine pollen rain is heavy enough to cover all parts of southeastern Arizona, including the Sonoran Desert; one may expect at least a fraction of a percent at any given point. Within metropolitan Phoenix, Walkington (175) found a fallout of 33 grains per square cm. in the year 1957-1958, higher than the fallout of mesquite. In addition to native species in the mountains some of the pollen may be derived from cultivated Aleppo pines.

In southern Arizona pinyon commonly occurs as low as 1,650 m., where it occupies the upper part of the oak-juniper woodland. In deep ravines or on especially favorable sites, both pinyon *(Pinus cembroides)* and Chihuahua pine *(P. leiophylla),* and even yellow

TABLE 1. MODERN POLLEN RAIN, PERCENT COUNT (N = 200)

	Source		County and Location	Regional Vegetation	Elevation (in meters)	Pinus	Quercus	Fraxinus-Juglans	Picea-Abies	Juniperus	Celtis	Prosopis	Acacia-Mimosa	Agave-Cactaceae	Ephedra	Σ AP	Cheno-Am	Compositae	Cheno-am/Comp ratio	Gramineae	Euphorbiaceae	Malvaceae	Nyctaginaceae	Tidestromia	Plantago-Kallstroemia	Cruciferae-Gilia	Geraniaceae-Cyperaceae	Others + Unknowns
1.	Stock Tank	E	Graham; 5 km S Safford	Sonoran Desert	940	1	2	J-½	½-2			2			1	9.2	48	25	66	7	1	½	½				G-1	9.6
2.	"	E	Pima; 5 km NW Helvetia	Shrub-Grassland	1,000	2	½				2	3	A-½	A-1		8.7	8	42	15	28	2			4				7.9
3.	"	R	Pinal; 2 km NE Oracle Jn.	"	1,040	1	½			3		2				7.0	34	24	58	19	4				P-1	G-½	G-3	7.9
4.	"	T	Pinal; 2 km NE Oracle Jn.	"	1,040	1	1				½	3			½	5.5	63	16	79	9	1		½	½		G-½	G-1	3.2
5.	"	E	Pima; 17 km SW Redington	"	1,210	2	1				1	6				10.2	18	35	34	24	5	1	½					7.4
6.	"	E	Cochise; 2 km W Double Adobe	Grassland-Shrub	1,220	2	4	J-½			1	3				10.7	32	9	78	38								9.8
7.	"	T	Cochise; 7 km E Double Adobe	"	1,225	1	1			½		½				2.8	54	6	89	25	4		2	2	K-½			3.2
8.	"	R	Cochise; 10 km NE Double Adobe	Grassland	1,260	1	3									4.2	75	3	96	9	6					C-½		2.3
9.	"	R	Cochise; 13 km NE Double Adobe	"	1,280		½									0.5	82	5	94	6	2		½			C-½	C-1	1.8
10.	"	R	Cochise; 9 km ESE McNeal	"	1,260	½	3									3.3	84	3	96	8								1.9
11.	"	R	Pima; 16 km S Pantano	Grassland-Shrub	1,270	3	1			1		1				6.8	53	4	92	33								2.9
12.	"	R	Cochise; 6 km SW Willcox	"	1,270	½	1	J-½		½		3	M-½	A-½		5.7	42	1	94	48							C-1	1.8
13.	"	E	Pima; 6 km W Benson	"	1,280	2	10	F-½			½	1			1	14.6	54	12	82	16		½		½	K-½		C-½	2.5
14.	"	R	Mexico-Chih.; 28 km S Antelope Wells	Grassland	1,380	4	3					3				9.8	56	17	77	12				5				1.9
15.	"	R	Mexico-Chih.; 35 km S Antelope Wells	"	1,380	6	6								1	13.1	56	9	87	15	½			1				4.7

16.	” R	P ma; 9 km N Sonoita	”	1,420	4	12	J-½	1	2		1	20.6	24	11	69	36		G-1		7.5
17.	R	Santa Cruz; 2 km E Sonoita	”	1,460	8	11		2	½			21.6	34	14	71	26				5.0
18.	” R	Santa Cruz; 8 km NW Sonoita	Oak-Grassland	1,520	13	15	J-2		½			30.4	38	7	85	10			C-½	11.8
19.	” T	Pima; 12 km W Twin Buttes	”	1,580	3	46		1	3	9		61.1	8	7	54	7	½	C-1		15.9
20.	” R	Graham; Point of Pines	Pine-Parkland	1,820	12	4		4	½			20.9	10	52	17	12				3.5
21.	” E	Graham; Point of Pines	”	1,880	53	19						71.0	5	14	26	6				4.5
22.	” E	Graham; Point of Pines	”	1,880	14	3						17.0	30	35	47	13				5.0
23.	Summer Flood	Pima; Santa Cruz R., Tucson	Sonoran Desert	670	½	½			1	1		2.8	47	32	13	½		P-1		4.6
24.	Summer Flood	Cochise; 9 km NW Bisbee	Shrub-Grassland	1,340	18	1	1-½	½	½	6	C-1	27.5	13	26	22	4	½	2		5.7
25.	Arroyo Bed	Cochise; 10 km N Double Adobe	Grassland-Shrub	1,230	2	½				5	½	8.1	55	15	19	½		1		2.9
26.	”	Cochise; 10 km N Double Adobe	”	1,230	1	1				3	½	5.9	55	19	18	½		P-1		2.5
27.	Cienega Soil	Pima; 14 km NNE Sonoita	”	1,330	3		½-1				½	5.1	68	14	8	1	½		½	1.5
28.	”	Graham; Point of Pines	Pine Parkland	1,880	32	4		½				36.1	17	23	16					9.8
29.	”	Graham; Point of Pines	”	1,880	17	6		6				28.8	8	21	20	½			C-3	12.5
30.	Spring	Graham; Point of Pines	”	1,880	32	10		1				43.1	8	28	13					5.5
31.	(6) Cow Dung	Cochise; 1 km W Double Adobe	Grassland-Shrub	1,210								0.0	78	½	17		1			2.0
32.	(11) ”	Pima; 16 km S Pantano	”	1,270	3	½				28	9	39.8	19	9	18	½				13.0
33.	(20) ”	Graham; Point of Pines	Pine Parkland	1,820	3	½						3.4	½	87	9			P-1		0
34.	Soil Surface	Pima; Catalina Mts.	Pine Forest	2,430	81	2						85.4	2	7	2					3.3

17

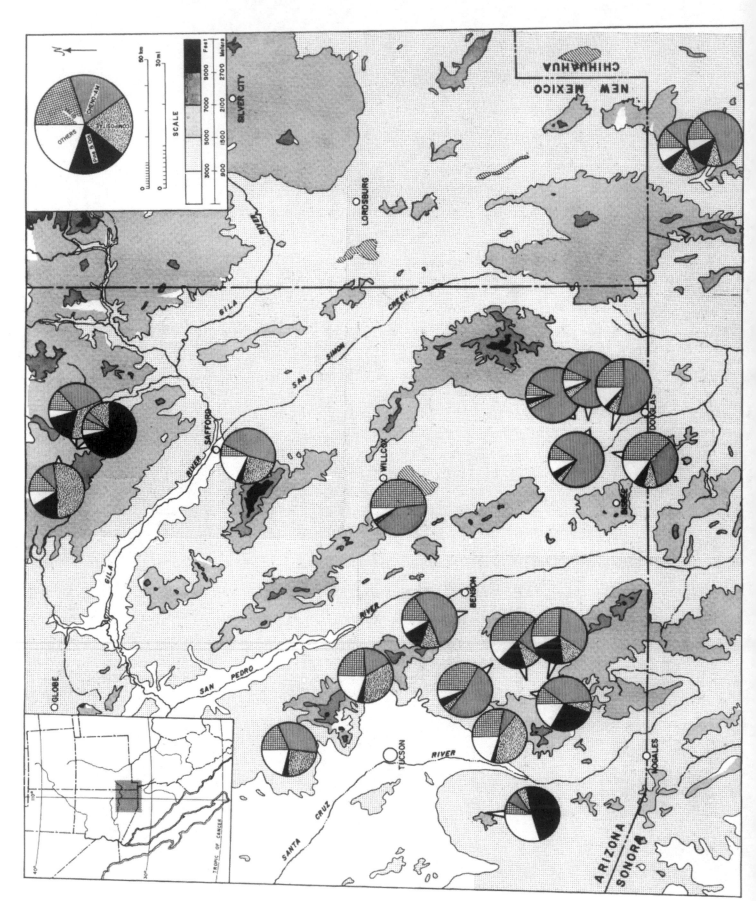

Figure 7. Pollen analysis of stock tanks in southern Arizona.

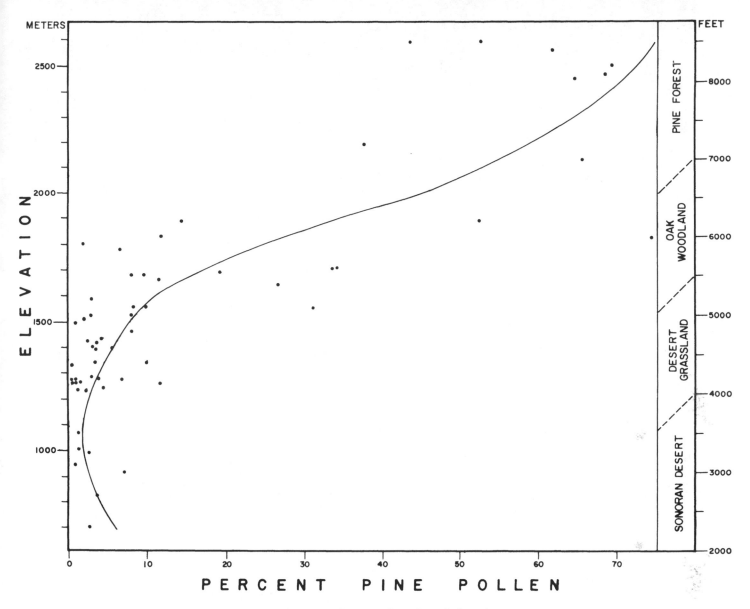

Figure 8. Pine pollen as a function of elevation.

pine, may descend to 1,500 m., as in Cave Creek of the Chiricahua Mountains. *P. leiophylla* occupies a *south slope* at 1,400 m. in the Patagonia hills (115). None of these low records of pine were reached near any of the cattle tank sampling stations. Above 1,500 m., in areas where pine is present or no more than three to four km. distant, the percent frequency of pine pollen is much higher than it is at lower elevations (Fig. 8), and in the Catalina Mountains at 2,250 m. forest soil beneath a canopy of *Pinus ponderosa* shows a pine pollen frequency of 70 percent. But even here low-spine Compositae and cheno-am pollen can be found in soil samples, despite absence of such plants in the vicinity. While pine travels long distances into the desert and desert grassland, there is also herb pollen transport from arid habitats into the mountain forests (112).

A second means of relating fossil pine pollen content to various natural communities is by size-frequency analysis. Differences in pine pollen size have been known at least since the time of Engelmann (58).

While measurements of fresh, acetolyzed material cannot be used to identify individual fossil pollen grains, I assume that the size-relationships seen in modern species (Fig. 9) provide a basis for interpreting relative size changes in the fossil record. Unless diagenetic changes have differentially altered size of fossil grains, it should be possible to derive both stratigraphic and paleoecological information from pine pollen measurements. The measurement I have adopted is bladder length. It can be obtained generally on broken as well as on perfectly preserved pollen grains. In both my experience and that of Wenner (178) body length is more variable and is difficult to obtain on broken or damaged grains.

One hundred pollen grains from male cones of a single individual of each of the nine Arizona species were measured by J. Schoenwetter, including within *Pinus ponderosa* two collections of *P. p. arizonica* from the Chiricahua Mountains and one of *P. p. scopulorum* from Prescott (Fig. 9). Bladder length (= bladder breadth of Wenner [178] and the saccus breadth 10-10

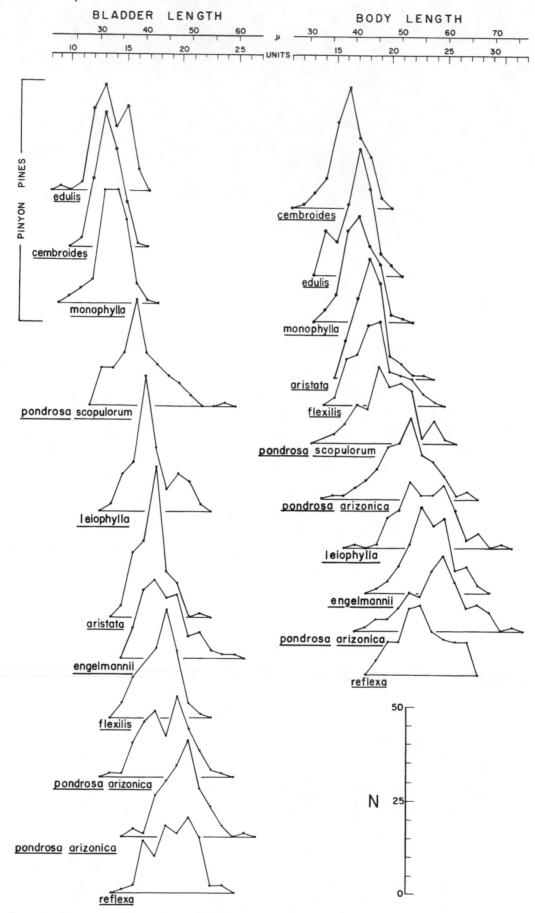

Figure 9. Pollen size-frequency histograms of Arizona pine species. Pinyons are the smallest. One hundred pine bladders were measured from one tree of each species.

measurement of Erdtman [60] and body length [corpus breadth 5-5 of Erdtman]) were recorded under 400x magnification. The smallest species are the pinyons, *Pinus edulis, cembroides,* and *monophylla.*

Ecologically it appears that many species of pines with small pollen size are also those which tend to occupy the more xeric sites (either cold-dry or warm-dry) which are unfavorable for rapid growth. In addition to the pinyons and jack pine, *P. engelmannii* and *P. aristata* are slow-growing species adapted to relatively extreme environments. They have relatively small pollen grains. Species of large pollen size are often those which occupy the more xeric sites (either cold-dry or warm-dry) insure rapid growth. *Pinus ponderosa arizonica* and *P. reflexa* fall in this category.

It is possible that populations within a single species might also respond to varied environmental conditions with a change in pollen size. Measurements of pollen size of *P. ponderosa* in the Catalina Mountains were taken throughout the range of the species, from 1,800 m. to the top of the mountain at 2,760 m. The lower elevation trees shed pollen a month in advance of those at the top. There was a slight indication of a maximum pollen size at 2,580 m., with smaller populations above and below (see Table 2). However, there is a large amount of variation between trees from a single elevation.

Under the assumption that size of bladder bears a rough relationship to site ecology and is an index of growth conditions, one may investigate past environmental change. As Engelmann (58) observed, pinyons provide the smallest size pollen (Fig. 9). Large size pollen in the fossil record should be derived from yellow pines or white pines. In ecological terms one may hope to distinguish woodland of pinyon-juniper or oak-pinyon-Chihuahua pine from parkland and forest dominated by *P. ponderosa* and *P. reflexa*. For example, the mean size of fossil pollen from alluvium in the desert grassland (Double Adobe I) is significantly smaller than pine pollen within yellow pine parkland at Point of Pines (Fig. 33).

Oak and other AP. Although oak is dominant in the woodland from 1,500 to 2,300 m., its pollen is usually less abundant than that of pine. An exception is the high frequency, 46 percent, of oak pollen from a tank west of Twin Buttes which lay in the shade of *Quercus emoryi*. Ordinarily oak pollen may reach its maximum abundance just above 1,500 m. (Fig. 10).

Of the high montane conifers *Abies* occurs on all the mountains of the Southwest which rise above 2,700 m. It is much more abundant than *Picea* which occurs south of the Mogollon Rim only in the Chiricahuas and Pinaleños. Pollen of both spread from the latter range into a tank at 940 m. in the Sonoran Desert near Safford. Alder (not shown on Table 1) also occurred in this tank. Obviously, distance to the nearest source of montane mesophytic trees will affect pollen fallout within the desert and grassland. Considering the prolific pollen production by conifers, their occasional appearance in stock tanks of the desert grassland is not surprising.

Although not a prolific pollen producer, mesquite *(Prosopis juliflora)* was encountered in 13 of 22 tank samples. Mesquite pollen was found in two samples of fresh alluvium from the bed of Whitewater Draw near the site of Double Adobe profile IV. The infrequent occurrence of mesquite in the upper 20 cm. of fine silt sedimented during historic time and largely prior to 1900 may provide further documentation of the recent mesquite invasion of certain areas, that is, after arroyo cutting. Earlier records of flood plain mesquite are seen in older sediments of various profiles, especially at Double Adobe II.

Despite local over-representation of NAP in mountain meadows and natural prairies at higher elevations, the arboreal pollen sum (AP) increases with elevation, as one would expect (Figs. 7 and 11).

Cheno-ams, Compositae, and other NAP. The ratio of cheno-ams to Compositae in stock tanks appears to change with altitude (Fig. 12). Cheno-ams reach a maximum between 1,200 and 1,300 m., where they may exceed 90 percent of the total pollen count. At lower elevations they are replaced by *Franseria, Hymenoclea,* and various high-spine Compositae pollen types. At higher elevations in southern Arizona broad valleys and alkaline soils are lacking, and the cheno-ams give way to low-and high-spine Compositae pollen.

Although cheno-ams predominate in most modern pollen samples from the desert grassland of southern Arizona, they are occasionally replaced by low-spine Compositae *(Ambrosia)*, as in the Empire Cienega. Four km. east of Empire Ranch the Empire Gulch widens

Table 2. Size frequency of **western yellow pine** pollen at different elevations in the Catalina Mountains. Male cones of individual trees were acetolyzed and 20 bladder measurements (longest axis) recorded in microns. Data shown are means, twice the standard error of mean, standard deviation, and range.

No. of Tree	Elevation in Meters	Mean $\pm \frac{2\sigma}{\sqrt{N}}$	σ	Range
1.	1,820	37.7±2.0	4.6	30-46
2.	"	40.4±1.6	3.5	36-47
3.	1,980	36.7±1.5	3.2	30-43
4.	"	41.5±1.6	3.5	35-46
5.	2,280	35.7±1.9	4.2	27-42
6.	"	37.6±1.7	3.4	32-48
7.	"	37.9±1.7	3.7	30-45
8.	"	38.6±1.9	4.4	30-45
9.	"	40.2±1.5	3.4	34-49
10.	2,580	40.4±1.8	4.1	33-50
11.	"	40.7±1.7	3.9	30-46
12.	"	41.5±1.7	3.7	33-49
13.	2,770	35.1±1.4	3.1	30-40
14.	"	37.2±1.5	3.3	30-47
15.	"	38.1±1.5	3.4	33-44
16.	"	38.2±1.5	3.3	31-44

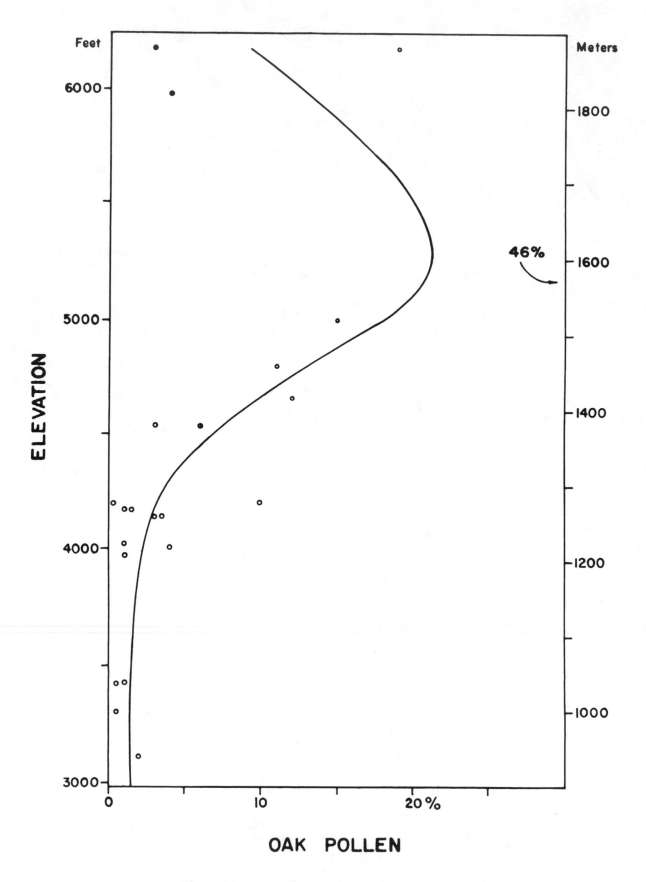

Figure 10. Oak pollen as a function of elevation.

Table 3. Surface pollen samples, Empire Cienega, Pima County. N=200. Ten meters separate each station. Pollen count based on analysis of a mixture of ten sub samples from a one meter plot. Percent plant cover in each plot shown on Fig. 13.

	Pinus	Quercus	Juniperus	Celtis	Prosopis	Salix	Dodonaea	ΣAP	Cheno-ams	Low-spine Compositae	High-spine Compositae	Liguliflorae / Artemisia	Gramineae	Euphorbiaceae	Eriogonum	Plantago	Nyctaginaceae	Leguminosae	Cyperaceae	Polemoniaceae	Evolvulus	Unknowns
1.	7	5	1		8	1		22	80	57	4	1	34		3							3
2.	2	1			5			8	125	38	7		15		7							4
3.	4	5			1			10	157	15	9		9									2
4.	4	2						6	164	6	7		14		3							1
5.	5	5						10	145	21	17								1			3
6.	2	1		1				4	182	5	4		3			1			1			
7.	1	3		1				5	156	17	6		14		1				1			3
8.	5	2						7	116	41	18		13		1				3	8		5
9.		8		4		2		14	86	45	15		5		1		1		2		1	10
10.	3	5		1			1	10	100	43	19	1	25					2				2
11.	5	3	1		1			10	50	107	13	1	17	1				1				4
12.	1	3			1			5	8	147	19		18	1	1				1			
13.	1	1			1			3	17	167	1	1	10		1							
14.		2			1			3	17	174			4	1			1					
15.	2	5						7	34	127	12		16	3					1			2
16.		5	1					6	54	114	9		17									1
17.	4	3						7	20	153	5		14	1								4
18.	4	4			1			9	67	86	3	1	30	1	1				2			2
19.	1	1			1			3	13	165	3	1	9	1					5			
20.	1	1	1		10			13	78	76	11	1	18				1		1			2
21.	1	2	1		20	2		26	65	39	9		58	3								4

into a level, unterraced flood plain, enclosed by gentle bajada slopes, and, unlike most flood plains of southern Arizona, it is uncut by an arroyo. The south edge of the Empire Cienega contains populations of both perennial ragweed *(Ambrosia psilostachya)* and giant ragweed *(A. trifida)*.

A transect across the Empire Cienega through the ragweed colonies revealed predominance of cheno-am pollen in soil associated with the local distribution of *Atriplex*. There was a predominance of low-spine Compositae pollen from under or near the patches of giant ragweed (Fig. 13, Table 3). On September 3, 1962, measurements of plant coverage were made at 21 stations, each a meter square and 10 meters apart across the cienega. Pollen was extracted from a mixed sample of 10 subsamples taken at each station. The relative abundance of cheno-am pollen and of low-spine Compositae pollen closely follows estimates of coverage of these plants; the relationship between pollen count and coverage of both grass and high-spine Compositae is not closely related.

It is likely that the perhistoric pollen zones domi-

nated by low-spine composite pollen represent the type of environment found in the Empire Cienega today — a relatively undissected flood plain with a high water table and non-saline soil, supporting dense colonies of giant ragweed.

Regarding the modern pollen record of grass in cattle tanks from the Southwest, there is an apparent drop in frequency in woodland and forest zones. Otherwise the grass frequency bears no clear relationship to altitude (Figs. 7 and 14). Stock tanks adjacent to dense flood plain swales of *Sporobolus* may yield 48 percent grass pollen. Those on the bajadas within the *Bouteloua-Aristida-Hilaria* forb-grassland may yield less than 10 percent Gramineae. Grasses vary greatly in pollen production (94) and the native upland species may be poor producers.

Concerning other NAP, little can be said at this point beyond noting that metal-rim tanks appear to contain as many pollen grains of the zoophilous Malvaceae, Nyctaginaceae, and Euphorbiaceae as do the dirt tanks. These plants may liberate more pollen into the air than other animal-pollinated species.

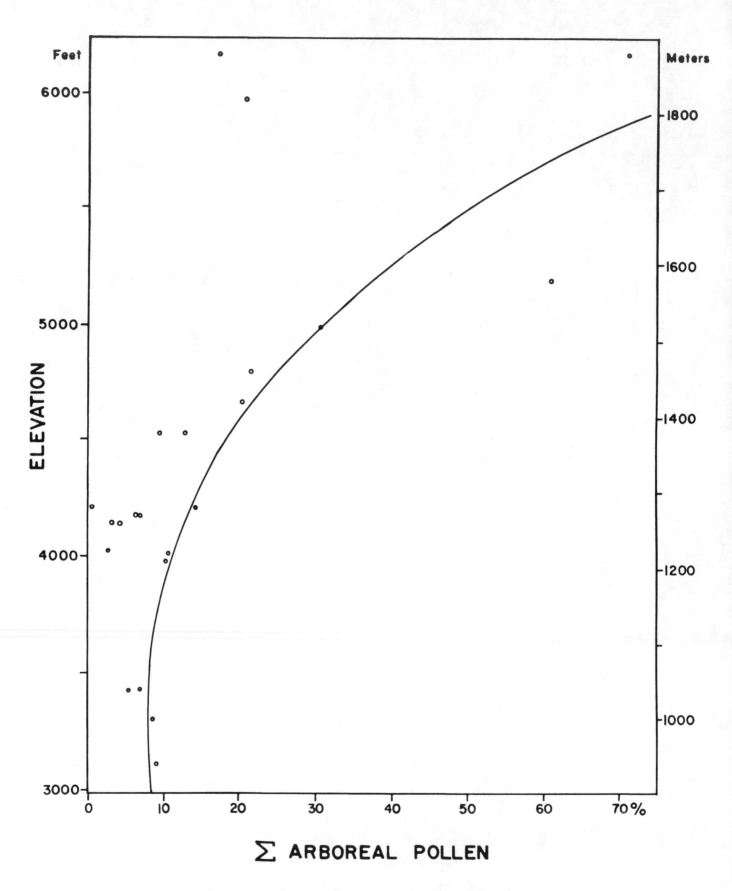

Figure 11. Aboreal pollen sum as a function of elevation.

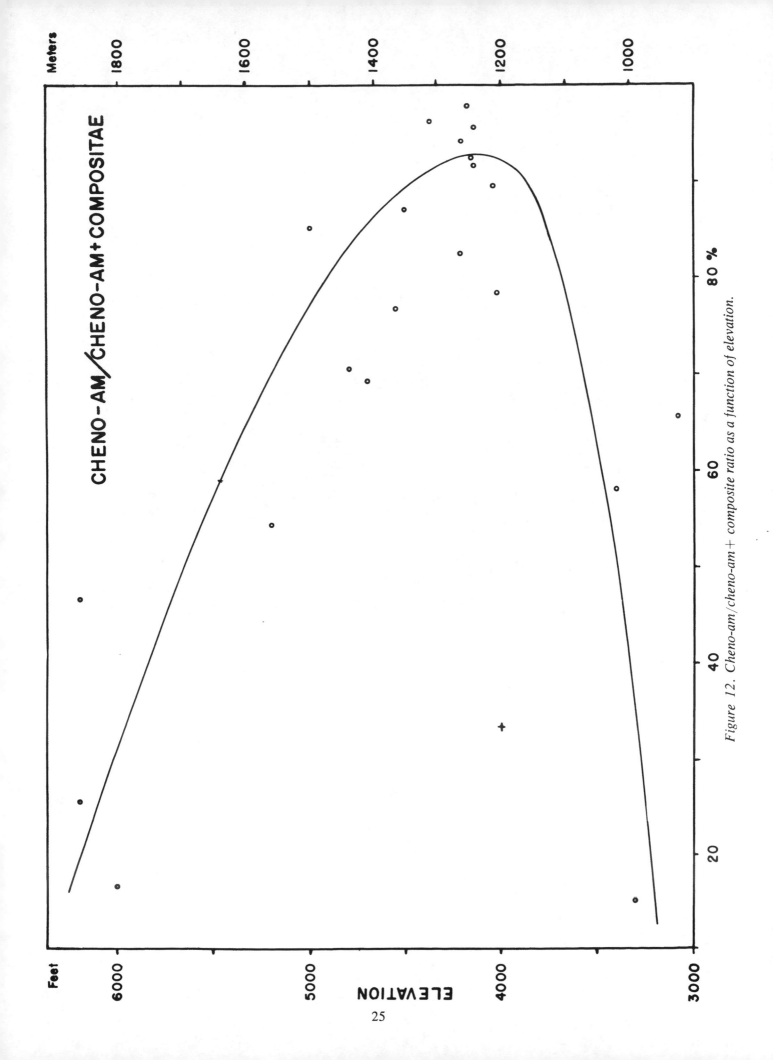

Figure 12. Cheno-am/cheno-am+ composite ratio as a function of elevation.

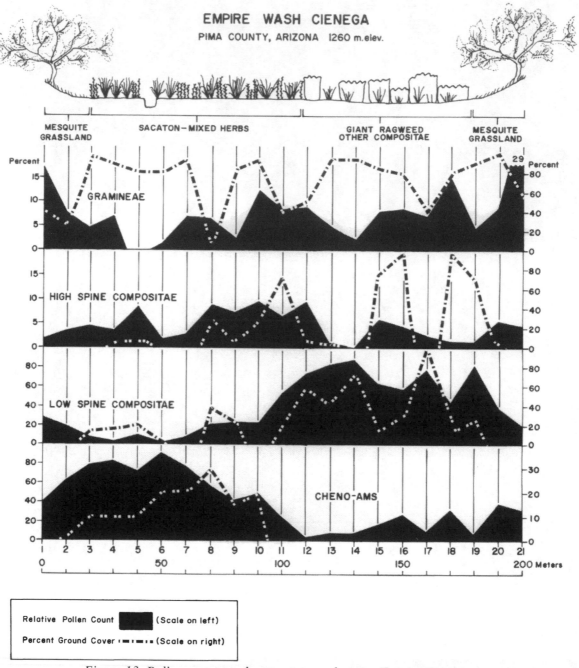

Figure 13. Pollen counts and percent ground cover, Empire Cienega.

Pollen Content of Extraneous Pollen Traps. A few pollen sources in addition to the stock tanks were sampled. Some, such as cow dung, may be of economic or biological interest, but because of seasonal variation, they present little immediate promise as a substitute for tank samples in the study of the local pollen rain.

Flood water scum, collected from Dixie Wash in August 1959 (No. 24 on Table 2), was remarkable in several features. The sample was collected by skimming handfuls of froth from runoff after a heavy thunderstorm in the Mule Mountains. The frequency of conifer pollen was unexpectedly high, and the grains were perfectly preserved. A few pinyon (*Pinus cembroides*) occur on the top of the Mule Mountains at about 2,370 m. (Photograph 5); these could scarcely account for

the presence of 18 percent pine in the count from the foot of the mountains below the limit of trees. Spruce and fir appeared in small numbers in the initial 200 grain count; a scan of the entire slide (N = 4,940) established the frequency of spruce as 0.45 percent (22 grains), that of fir as 0.18 percent (9 grains). The nearest present source of these is in the Chiricahua Mountains, 65 km. to the east.

Presence of so many conifers and their excellent state of preservation may represent selective concentration and floatation, perhaps similar to the shore drift or "seebluten" of estuaries (59, 61). Should the process occur on a large scale it could operate to reduce the conifer pollen content in alluvial beds, with the froth-trapped conifer pollen grains lodged and eventually

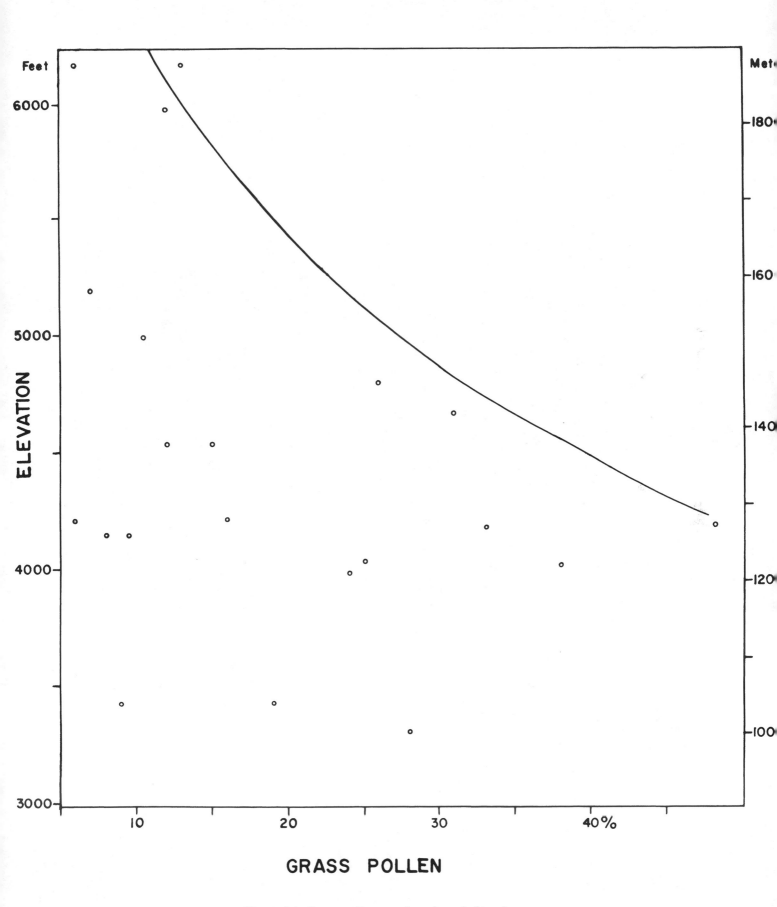

Figure 14. Grass pollen as a function of elevation.

oxidized in piles of driftwood above ground. The fresh alluvium deposited in the Whitewater Draw in the summer of 1959 (samples 25 and 26 on Table 1) bears no resemblance to the froth collected on the tributary Dixie Wash.

The grass pollen in the latter sample was abnormal. Of 100 grains examined closely, 17 percent showed two pores and 2 percent had three pores. The abnormality might reflect polyploidy in adjacent grass populations. One fossil diporate grass was observed in pollen zone IV of the Double Adobe I profile.

A direct estimate of the modern pollen rain can be obtained from the upper one cm. of the desert soil. Well-preserved pollen occurs in all soils in the Southwest, from the Sonoran Desert to fir forest at 2,700 m. on Mount Lemmon. In view of Dimbleby's observation (52) that pollen is rapidly destroyed in basic soils, the discovery of well-preserved pollen in the alkaline soils of Southwestern deserts and grassland was not anticipated. The seasons of microbiological activity in this area are perhaps too short to destroy pollen rapidly.

For want of a more appropriate place, I will comment here on the matter of pollen burial by insects, and the remote possibility that such pollen might contaminate the flood plain record. Of eight solitary bee nests collected by Dr. K. V. Krombein from borings in wood near Portal, Arizona, four contained only mesquite pollen (N = 14,000) and another was 96 percent mesquite. Of the others, one contained *Opuntia* pollen, another an unidentified Papillionoideae, and the third an unknown, possibly of the family Scrophulariaceae.

Many solitary bees burrow into the ground and deposit pollen at the end of a tunnel. Some prefer to nest in sandy soil (107). It is possible that a nest, or its remains, might inadvertently be included occasionally in a pollen sample of fossil alluvium.

V. THE FOSSIL POLLEN RECORD

Source and content of the profiles. My brief review of recent advances in Pleistocene geology, climatology, and vegetation analysis, and my sketch of the natural pollen rain of undisturbed plant communities in the arid Southwest, must now be put to the test. Will they provide sufficient insight to account for changes in the fossil pollen record of the flood plains? Of one thing we can be certain. Except for the Willcox Playa profile, man was on the scene. The pollen diagrams will, hopefully, reveal the nature of the vegetation and climate which prehistoric man experienced.

Of the 13 pollen diagrams, 11 were collected at localities in the desert grassland of southern Arizona, 1,200 to 1,350 m. elevation. The remaining two came from higher elevation near Point of Pines in parkland of yellow pine, oak, juniper, and pinyon. All except the Willcox Playa and the Turkey Creek site represent

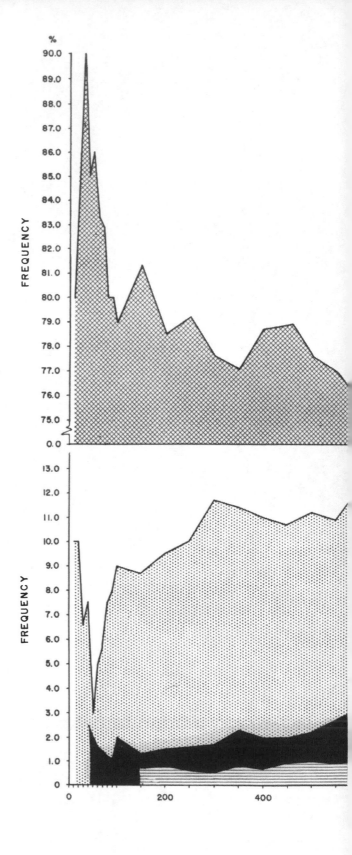

alluvial deposits. Floods of the last 80 years have exposed ancient alluvial beds, and in most cases pollen samples were collected from the freshly cleaned vertical face of an arroyo, often with the help of a ladder to reach the upper levels. Through inspection of the arroyo walls it is possible to describe bedding, erosion surfaces, position of artifacts, and to locate appropriate samples for radiocarbon dating. In this respect the study of

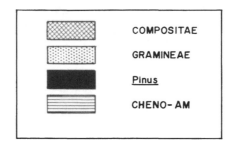

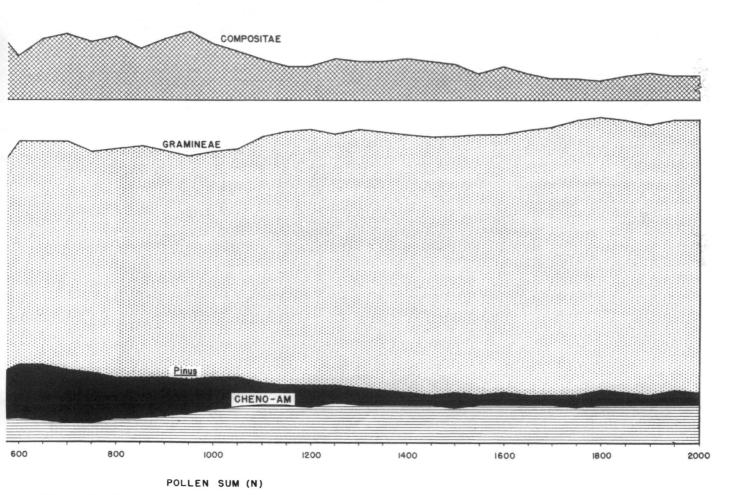

Figure 15. Change in frequency as a function of sample size.

arroyo stratigraphy has distinct advantages over the investigation of lake sediments.

To benefit from previous field studies it seemed prudent to begin with pollen samples collected at localities of established archaeological or biological significance. Such a locality is Double Adobe, Cochise County, Arizona. In their pioneering study of the Cochise Culture and flood plain alluvium, Sayles and Antevs (151)

demonstrated that three pre-pottery cultural stages could be detected along the Whitewater Draw. Recently they have added a fourth (152). In order of increasing age these stages are as follows: (1) San Pedro — pithouses, storage pits, deep-basin metates and large manos; (2) Chiricahua — no pithouses, basin metates, shaped manos, percussion flaked tools and pressure-flaked points; (3) Cazador — moderately large basin metates, small shaped manos, pressure-flaked projectile points; (4) Sulphur Spring — flat-surface metates with natural pebbles for manos, no projectile points. The Sulphur Spring artifacts were encountered with or below bones of extinct mammoth, camel, horse, and dire wolf.

Eight pollen profiles (Double Adobe I, II, III, and IV, Murray Springs, Cienega Creek — Empire Valley, Matty Wash, and Cienega Creek site — Point of Pines) are associated with sediments enclosing Cochise Culture remains, including the type sites of Sulphur Spring and Cazador stages. Three profiles can be related to the late Pleistocene megafauna (Double Adobe I, Lehner site, Malpais site). Ten profiles are adjacent to or within strata dated by radiocarbon samples (Double Adobe I, II, III, and IV, Lehner site, Cienega Creek, Matty Wash, Murray Springs, Willcox Playa, Cienega Creek site — Point of Pines, see Table 6). The result of correlating diagrams from the desert grassland is shown in Fig. 36.

Extraction and Counting. Laboratory extraction followed standard methods for treating inorganic sediment, including both tetrabromethane floatation and hydrofluoric acid digestion followed by acetolysis (61). In many cases standard treatment led to formation of a stubborn silicate colloid which obscured the pollen. To overcome this problem, and to reduce the volume of sediment needed for each run, a method of preconcentration was used (9). Ground, untreated sediment is frothed in a solution of pine oil, detergent, and quebracho. The latter acts as a silicate depressant and a slow steady evolution of bubbles will float away non-siliceous material, including pollen, into a collecting beaker.

The preconcentration is a fractional process; a certain amount of pollen is not separated from the sediment, and much inorganic material also floats up in the froth. The method made it possible to count certain strata which were poor in pollen, or likely to form a dense colloid when not preconcentrated.

The possibility of artificial concentration of certain pollen types was tested in several ways. Conceivably preconcentration might increase the amount of buoyant grains, as pine. It might fail to float mineralized pollen. In each of three profiles paired samples were analyzed to compare results with and without pretreatment.

The first was of the Point of Pines Cienega site, in which the lower bar at each level represents the sample subjected to preconcentration, the upper represents the sample extracted by conventional methods (Fig. 29). On the assumption that percentage changes in pollen counts can be compared using confidence intervals based on the binomial distribution, it is obvious that significant differences (at the 95 percent confidence level) occur between many replicates.

In all profiles a sample size of 200-250 was used. With an N of 250 and for percentages between 15 and 85; the 95 percent confidence level extends five to seven percent on either side of the observed value. For ob-

served values larger than 85 percent or smaller than 15 percent the confidence level diminishes rapidly. Five of 15 paired samples within the Compositae curve, three of 14 levels in the pine curve, and eight of 13 levels in the sedge curve exceed what might be expected on the basis of chance alone. In the case of the Compositae and pine, no consistent trend is evident; if anything there is a tendency for pine to be less abundant in the preconcentrated samples.

On the other hand, sedge pollen is consistently more numerous in preconcentrated samples, a feature that is difficult to understand in terms of an extraction effect. As an alternative, I suggest it may reflect poor staining in the initial count; sedge is easily overlooked if the sample is not well stained and if the background is heavy.

In a second replicate analysis, Double Adobe II (Fig. 22), the initial count of Schoenwetter on samples treated routinely (lower bars) was followed by one by Martin on samples that had been preconcentrated (upper bars). Again an increase in sedge is evident. In general the Compositae curve shows consisent results. A tendency toward higher percentage of cheno-ams and lower percentage of pines in the sample not subjected to preconcentration is difficult to explain. It could mean that the preconcentration step selects pine differentially. On the other hand the high pine frequency in pollen unit 1 of this profile does not resemble the equivalent unit of the adjacent profile, Double Adobe I, which was also preconcentrated. The upper levels of Double Adobe IV were counted twice by Schoenwetter; upper bars are counts of preconcentrated samples. Recovery of initial results was generally satisfactory (Fig. 19).

If the preconcentrated samples are responsible for an error in counts it may be in an overabundance of sedges. Overabundance of pine cannot be discounted, but did not appear consistently.

The Problem of Sample Size. The basis for the choice of sample size, roughly 200 grains per level, may deserve comment. Except for occasional scans of slides for pollen grains of unusual interest (for example, *Zea* and *Carya*) sample size was held to roughly 200 grains. While sufficient for estimating frequency of abundant pollen types, the 200 grain samples are too small to necessarily include rare items of ecological importance or to provide a meaningful quantitative record of pollen types of low frequency (less than two percent).

To determine how percentages might change with increasing sample size, Double Adobe II level 320 cm. was counted to a total sample of 2,000 pollen grains. Major oscillations in percent estimate damp off at N = 200. The difference between counts based on samples of N = 200 and N = 2,000 did not exceed four percent (Fig. 15). There was no tendency for large pollen grains (pine) to shift in frequency with increased sample size, as Dimbleby (52) had found in the case of Ericaceae tetrads. Therefore, it appears that

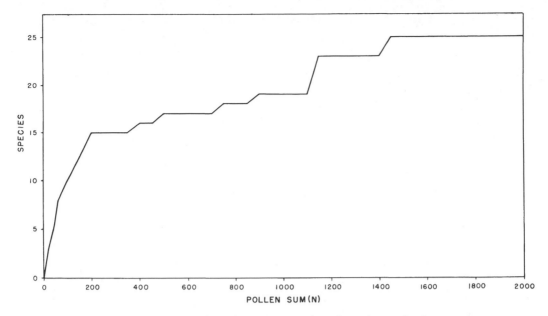

Figure 16. Number of species as a function of sample size.

little is to be gained in estimating the frequency of the common pollen types by adopting a larger sample size.

While the time-consuming practice of some analysts of counting all pollen on a slide may overcome the problem of non-random distribution under the cover slip, it usually results in a pollen diagram with very variable N, from perhaps 50 to 1,000 or more. Percentages based on samples of less than 150 will be subject to serious sampling error. Visual comparison of percentages derived from a small sample (less than 150) with a large one (1,000 or more) is hazardous. Finally, for purposes of multivariate statistical analysis, a fixed N of *exactly* 200 (or other size) is highly desirable.

By increasing sample size from 200 to 2,000 ten new pollen types were added (Fig. 16), an average of 3.0 per doubling of the sample. From studies of species of animals (145) there are reasons to believe the mean number of additions will remain constant through at least 15 doublings (octaves). If Preston's analysis also applies to pollen type distributions, increasing the sample size of Double Adobe II, level 320 cm. from 200 to a count of one million grains should add about 37 pollen types. In such terms one can gauge cost (observation time) of adding information (new pollen types).

The following abbreviations are used on the diagrams: Cheno-am = Chenopodiaceae plus *Amaranthus,* Euphorb = Euphorbiaceae, Mal = Malvaceae, Nyct = Nyctaginaceae, Cyp = Cyperaceae, Pol = Polemoniaceae, Onag = Onagraceae, Tid = *Tidestromia,* Kal = *Kallstroemia,* Art = *Artemisia.* The juniper curve may include *Cupressus.* Depending on relative abundance of the items illustrated, either a large or small scale was used. Numbers written to the left of the Compositae curve represent the count of Liguliflorae. For most pollen diagrams the exact pollen count of each pollen type

at each level is written to the right of each point on the curves. If radiocarbon samples were collected in direct association with pollen samples the dates are written with the sediment signature at the left margin of the diagram.

Within each diagram I have established stratigraphic units on the basis of major shifts in pollen composition. The pollen units thus established in each diagram were correlated into a system of zones (Fig. 36) based on pollen composition plus other available information, including archaeological stratigraphy, radiocarbon dates, and vertebrate fauna.

Pollen Profiles from the Desert Grassland

Cienega Creek, Empire Valley, Fig. 17. The Cienega Creek is a tributary of Rillito Creek and the Santa Cruz River, flowing north through the grass-swept Empire Valley, about 100 km. southeast of Tucson. Pollen samples were collected in the north end of the valley 20 km. north northeast of Sonoita on the west bank of Cienega Creek, elevation 1,280 meters. The exact point (Photograph 11) is about 20 meters upstream of Eddy and Cooley's measured section MC-6 (55).

Eddy's archaeological sequence begins with the preceramic San Pedro stage remains, including pithouses. At higher levels are various phases of the ceramic Hohokam period. The oldest radiocarbon date on flood plain sediments in the region is 3,570 ± 110. Early post-pluvial deposits comparable to the Sulphur Spring stage deposits at Double Adobe are unknown.

The flood plain sediments containing late post-pluvial fossil remains overlie a red calichified bajada surface which is exposed in the bottom of the present arroyo at certain points. The material comprising the

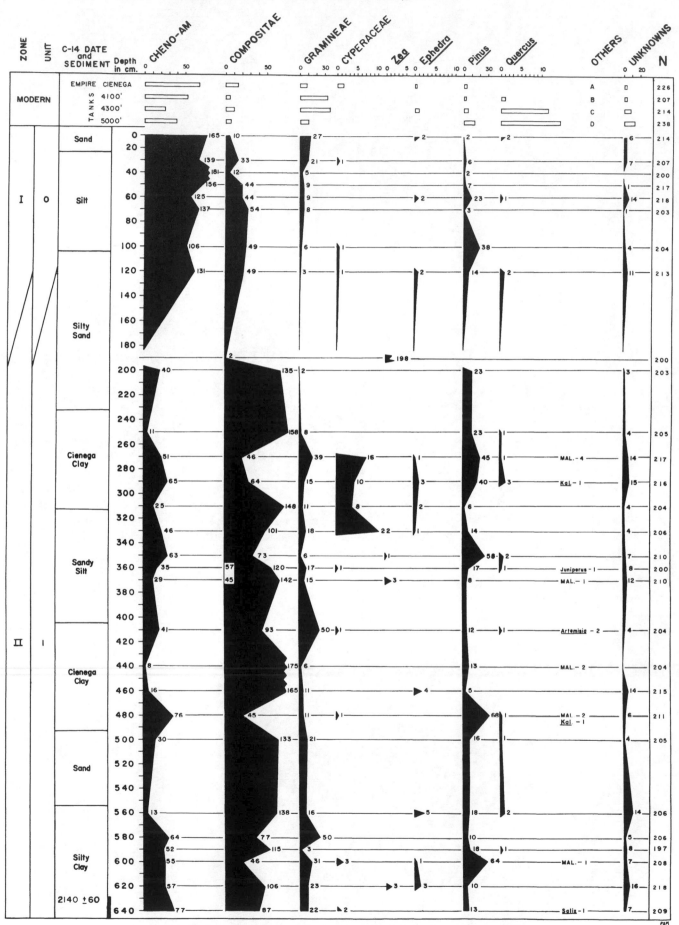

Figure 17. Pollen diagram of Cienega Creek, Empire Valley, Pima County.

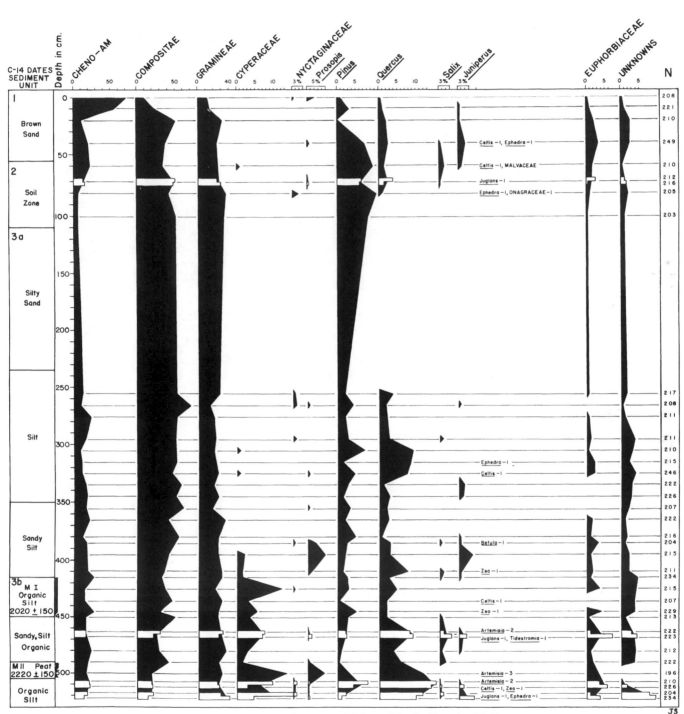

Figure 18. Pollen diagram of Matty Wash, Empire Valley.

top of the older valley fill is without fossils, either pollen or vertebrates. Its age is unknown; evidently it is much older than the recent flood plain sediments.

In the Cienega Creek flood plain deposits there are beds of very dark fine-grained silt. These may contain sedge pollen (levels 260-330 on Fig. 17); otherwise they are indistinguishable in pollen content from interbedded light sandy silts. I conclude that the dark cienega (= marsh) soils represent local ponding and are not diagnostic of regional climatic change.

The major change in pollen composition at the Cienega Creek locality occurs between levels 200 and 120 cm. where the dominant Compositae (zone II) are replaced by cheno-ams (zone I). This may represent a change in vegetation induced by cutting in the 13th century. Conspicuous channels were cut and filled into beds of late Hohokam age (55).

Along Matty Wash, a tributary of Cienega Creek and within half a mile of the Cienega Creek locality, Schoenwetter (153) encountered a similar shift from Compositae to cheno-ams at a position much higher in the section, within historic time. If the dating is reliable for both sites, an overall climatic control of the composite/cheno-am shift is ruled out and some other factor, perhaps differences in cutting at different points, may explain the change.

Of interest to the prehistorian is the presence of *Zea*. Curiously, only one other locality, the Point of Pines Cienega Creek site, exceeded the Cienega Creek locality in frequency of corn pollen. In other flood plain deposits, *Zea* was either absent or very rare (see Table 5).

A comparison was made between the alluvial flood plain sediments and recent stock tank samples near the Cienega Creek. Three stock tank samples from various elevations witihin the Empire Valley are shown on the Cienega Creek diagram (at the top of Fig. 17). The tank at 1,250 m. was within one kilometer of the fossil locality. Except for a significantly higher frequency of grass pollen, there is a close resemblance between upper levels of the flood plain and the cattle tank. A sample from the surface of the Empire Cienega, several kilometers south of the fossil locality, is also similar in pollen composition to that of the upper levels of the fossil horizon. Stock tanks at higher elevation show an increase in oak pollen and a decrease in frequency of cheno-ams.

Matty Wash, Empire Valley, Fig. 18. Alluvium exposed in Matty Wash, a tributary of Cienega Creek rising in the Whetstone Mountains, contains San Pedro and younger artifacts studied by Eddy (55). Close to Eddy's excavations and within a kilometer of the Cienega Creek profile at Eddy and Cooley's measured section MC-5, Schoenwetter (153) collected a pollen profile.

Exposed near the bottom of the west wall of Matty Wash at this point are two very dark humified beds containing fragments of plant material mixed with silt.

A radiocarbon date of the lower bed, M II (490-500 cm.), is 2,150 ± 70. Dates on the upper bed analyzed by different laboratories are S-5359, 1,850 ± 80 and A-88, 2,010 ± 150. These and other dates obtained from the Cienega Creek-Matty Wash alluvium indicate that none of the deposits are older than about 4,000 years. Absence of older deposits implies an intensive erosion, presumably during or immediately after the altithermal.

The Matty Wash diagram includes a high frequency of sedge pollen accompanying the organic "peat" beds. *Zea* is present near the bottom of the profile with the sedge. The decline of sedge pollen above four meters reflects a drying out of what must have been a marsh and its replacement with a drier cienega dominated by Compositae (probably *Ambrosia*) and grasses (probably sacaton). Although geologists often attribute a more humid climate to humic layers buried in alluvium, the pollen profile at Matty Wash does not reveal climatic change.

Cheno-ams are not dominant below the very top of the section; it appears that the top of pollen zone II is younger in the Matty Wash profile than it is at the Cienega Creek profile. Erosion may have begun earlier along Cienega Creek, leaving Matty Wash with its undissected flood plain "hanging" above Cienega Creek as is the cienega of the Empire Wash at present.

The main feature of both the Cienega Creek and Matty Wash diagrams is the long interval of Compositae dominance, a condition to be expected only in cienegas of undissected valleys with a rank growth of giant ragweed (*Ambrosia trifida*) and perennial ragweed (*Ambrosia psilostachya*).

Double Adobe IV, San Pedro Stage, Fig. 19. An interval of composite dominance in late postglacial time, perhaps comparable to that found in the Empire Valley was encountered in this section. Pollen samples were collected through fill of a late Cochise feature, perhaps a San Pedro stage pithouse or storage pit and below the floor to a total depth of 3.2 m. The locality is on the east bank of the Whitewater Draw, about 500 m. south of the El Paso gas pipeline crossing and about eight kilometers north northwest of Double Adobe, T. 22 S., R. 26 E., sec. 8, elevation 1,215 m. The fill was bisected by the present arroyo.

Replicate pollen counts were made to a depth of 110 cm. The upper count is that of Schoenwetter on material prepared by standard acetolysis, floatation, and hydrofluoric acid treatment. The lower count, also by Schoenwetter, is of material processed with a preconcentration in addition to the previous steps. Below level 130 the counts are entirely those of Martin.

The zone of composite dominance occurs between levels 60 and 110 cm., at and above a radiocarbon date of 3,860 ± 200, thus it is somewhat older than zone II in the Empire Valley. Below 120 cm. composites decrease while cheno-ams, pine, and grass increase (unit

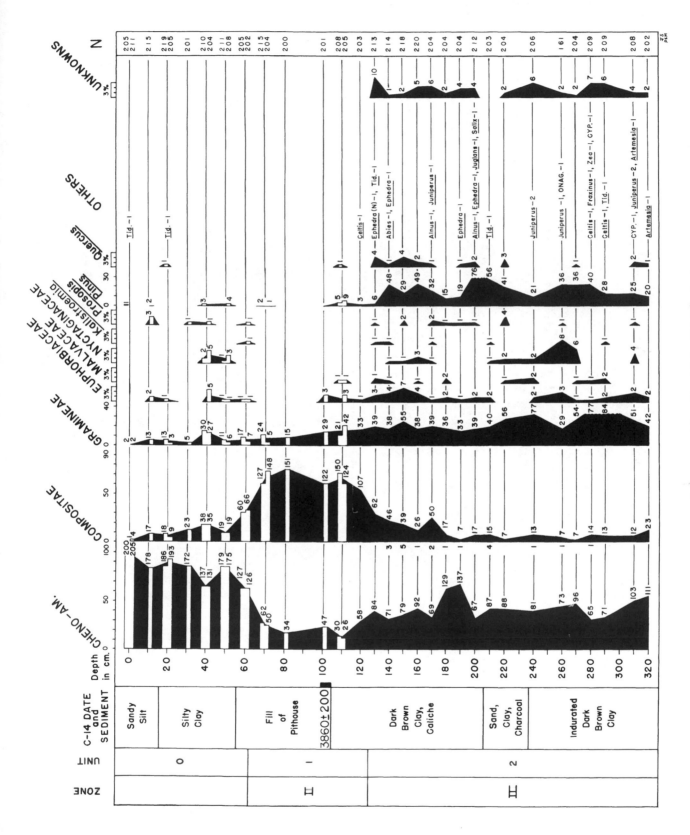

Figure 19. Pollen diagram of Double Adobe IV, San Pedro stage.

2 on Fig. 19). This unit bears a close resemblance to unit 1 of the pollen diagram of Double Adobe I; however, it lacks the hickory and elm pollen found at Double Adobe I, while the presence of *Zea* at 280 cm. suggests a younger age for unit 2 at Double Adobe IV.

The relatively high frequency of pine and grass pollen and the presence of a maximum of Malvaceae, Nyctaginaceae and *Kallstroemia* (all summer annuals in the desert grassland) indicates deposition during an interval of abundant summer rainfall. From its position beneath a radiocarbon date (A-193) which marks the end of the altithermal, I view the assembled pollen record as evidence of a warm, wet, subpluvial climate in what should be the altithermal. A date of $2,860 \pm 440$ from 235-245 cm. (A-194) is clearly discordant with my interpretation but A-194 is of carbonaceous alluvium rather than charcoal and thus it may be considered less reliable than A-193.

Left for last is the vexing matter of correlating pollen unit 0. It would be convenient to equate this to the cheno-am maximum in the loose silt at Double Adobe I and to the upper meter of the Cienega Creek profile, that is to pollen zone I as was done in Martin et al. 1961. However, the silty clay between 20 and 50 cm. is more thoroughly indurated than most alluvial deposits of the last 1,000 years. Loose silt at the top of the profile is part of a cavernous rodent burrow and mound. Furthermore, the abundance of stone artifacts including manos, metates, and scrapers in the immediate vicinity of the site suggests a late Cochise occupation of the unit 0 surface. In brief, the evidence for deposition of unit 0 within the last 1,000 years is not convincing, and no zone correlation is shown on Fig. 19.

Double Adobe I, Sulphur Spring Type Site, Fig. 20. In March 1959 two pollen sections were collected from the freshly cleaned face of the south bank of Whitewater Draw at Double Adobe. The sections lay 14 and 32 m. east of remnants of the bridge, now washed away, which Sayles and Antevs (151) used as a landmark for their initial excavations. The locality is 200 m. west of Double Adobe, elevation 1,200 m. The original Double Adobe Sulphur Spring site (Sonora F:10:1) lay immediately to the west of the old bridge. Traces of the old bridge are fast disappearing. Fill of its east pier lies 92.7 m. northwest of the west edge of the present bridge.

With the aid of a bulldozer a fresh alluvial exposure was cut to the top of the "pink caliche." The section was inspected by Sayles and compared with that previously excavated by Sayles and Antevs (152). While some uncertainty exists about the deeper beds, the general sequence of sediments appears to match their description. Unfortunately no artifacts were recovered in place which might establish with certainty the archaeological content of the beds. Early Cochise type artifacts were scattered in the backfill of the bulldozer cut (E. B. Sayles, pers. comm.).

Older sediments at this locality, including those of the Sulphur Spring stage, were considered to be a pluvial Wisconsin age deposit by Antevs. "The bones of the extinct animals, the much greater moisture indicated by the hickory, the permanent river, and the lakelet show that the sand-gravel and the laminated clay, beds b and c, were deposited during the last Pluvial, the correlative of the last glaciation in western North America" (151).

The upper pollen section from 20 to 500 cm. was collected on the south wall of the Whitewater Draw arroyo 16 m. east of the old bridge described by Sayles and Antevs (Photographs 12 and 13). The lower pollen section, at depth levels 10 through 100, was collected on the north wall to the east of the first set. The stratigraphic relationship of these two sections is shown in Fig. 21.

Sediment between levels 280 and 500 on the south wall was cross-bedded sand. Extraction of five levels from the sand indicated poor preservation or no pollen, and further analysis of them was not attempted. Sulphur Spring stage sediments appear to be represented in the upper section between levels 200 and 280 cm. In the north wall profile, I believe the entire section is correlative with the Sulphur Spring sediments. Admittedly, the present excavation long post-dated field work by Sayles and Antevs, and subsequent erosion had removed part of the beds they studied. Pollen samples collected by them in 1937 and sent to Sears cannot be located at this time. Nevertheless, I am confident that the sequence of pollen samples includes their beds b and c, as well as younger deposits studied by Sayles and Antevs (151).

Unit 0, from surface to 30 cm., dominated by cheno-ams, is a loose silt. In it Sayles and Antevs found sherds of 1300 to 1400 A. D. in age. Presumably the silt represents alluviation of the last 600-700 years.

Below the silt is indurated clay resting on a conspicuous erosion surface at about 120 cm. From 120 to 190 cm. is a white clay, highly calichified. Two of the calichified levels yielded no pollen (see Fig. 21). In the rest pollen was not abundant but was well preserved. The decrease in percentages of cheno-ams and the presence of a few grains of elm and hickory are the basis for the recognition of unit 1. Between 200 and 230 cm., cheno-ams become quite scarce; composites increase along with pine. The presence of *Polygonum* and sedges reflects local marshy conditions (unit 2). This unit is probably equivalent to the sediments described by Sayles and Antevs as laminated. Unit 3, below, is dominated by Compositae, which reach an abundance of 70 to 80 percent. The north wall, equivalent stratigraphically to unit 3 in the south wall, is quite similar but contains more *Fraxinus* and *Salix*. Pine is relatively scarce in both.

Both sections overlie a distinctive calichified valley fill, the pink clay ("bedrock") of Sayles and Antevs, seen to advantage only when freshly exposed. Resting

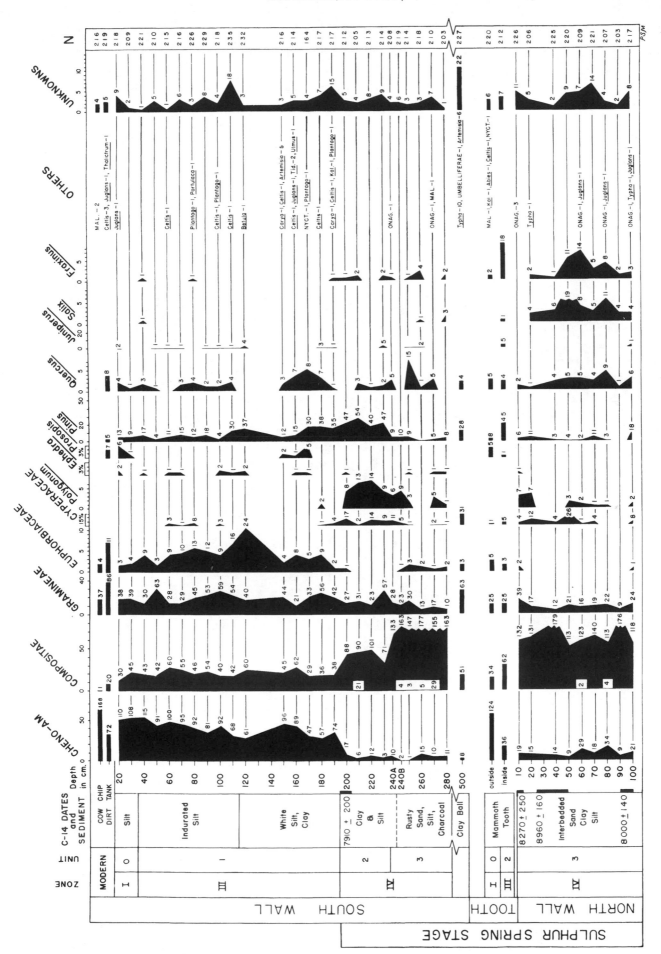

Figure 20. Pollen diagram of Double Adobe I, Sulphur Spring type site.

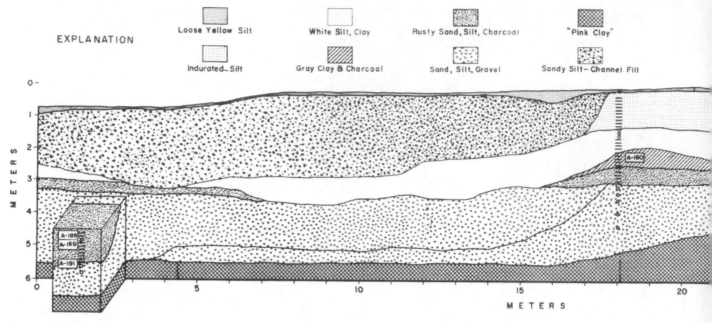

Figure 21. Profile of sediments and location of pollen samples at Sulphur Spring stage type site, Double Adobe.

on top of the valley fill and beneath crossbedded sand are occasional lumps of fine, plastic, gray clay. The analysis of one of these collected at a depth of 500 cm. showed a high frequency of sedge and grass pollen. The presence of *Typha* and sedge strongly suggests that the clay ball sample came from a pond deposit upstream. Although it is the oldest alluvial sediment analyzed at Double Adobe I, the clay ball lacks the high pine pollen count of the pluvial age core from the Willcox Playa (Fig. 28).

The Playa core samples contain much more tree pollen and *Artemisia* than any of the sediments from Double Adobe. The Playa samples show little resemblance to the alleged "pluvial" deposits at Double Adobe. If the Sulphur Spring sediments were of pluvial age they should show a much higher frequency of pine pollen, and radiocarbon dates from the beds should exceed 12,000 B.P. The combined palynological and radiocarbon evidence indicates an early postpluvial age, 8,000-9,000 B.P.

To determine whether the pollen profile would "date" remains of the Cummings mammoth collected in 1926, I analyzed the matrix from a tooth in the Arizona State Museum. The pollen spectrum obtained from mud inside lamellar plates of the tooth (See Fig. 20) is dominated by composites, but contains a fair amount of cheno-am and pine pollen. Except for a high amount of *Fraxinus,* this sample agrees quite closely with the average of the south wall spectra at 190 and 200 cm., approximately the depth below ground level at which the bones were found. It also resembles the percentage counts between levels 210 and 240 cm. at Double Adobe II.

Matrix from the outside of the tooth presents a totally different pollen spectrum. The abundance of cheno-ams, the very low amount of pine pollen, and the presence of mesquite suggest that the outer matrix is contamination from levels much higher in the section, presumably from unit 0.

Double Adobe II, Cazador Type Site, Fig. 22. A section was collected 108 m. west (north 83° west) of the old bridge and 300 m. west southwest of Double Adobe corners, elevation 1,200 m. Cultural remains excavated at this point are considered younger than the Sulphur Spring stage and older than the Chiricahua stage (152). Mr. Sayles helped to relocate a spot suitable for facing and sampling in the section which he had excavated previously.

Changes within the pollen diagram resemble closely those seen along the Double Adobe I south wall section. Pollen unit 0, dominated by cheno-am pollen, presumably is younger than 1200 A. D. It occurs to a depth of 50 cm. A rise in pine pollen can be seen and was used to distinguish units 0 and 1. Significant differences in pine frequency appeared between counts by Schoenwetter and Martin, but even the lower count by Martin shows unusually high percentages.

A white caliche bed found between 100 and 230 cm. can be traced to the white clay seen at Double Adobe I. Cheno-ams predominate in the caliche bed here as they do at the Double Adobe Sulphur Spring site. A single grain of elm (*Ulmus*) was found. Mesquite pollen is more common than in buried alluvium elsewhere. With increasing depth, Compositae and pine pollen increase in frequency; cheno-ams decline and mesquite disappears. Except for the absence or scarcity of sedge and *Polygonum* there is an obvious resemblance to unit 2 at Double Adobe I.

At the bottom of the profile composite pollen is very abundant; there is a relatively high frequency of Liguliflorae and pine is scarce. Such features also typify

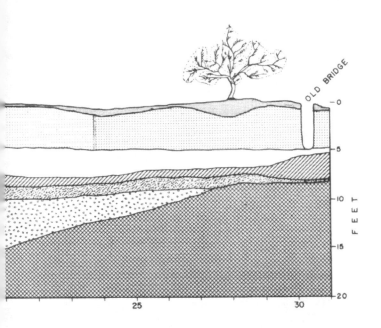

unit 3 at Double Adobe I. I conclude that these sediments are of the same age. A pollen count from pink calichified valley fill was obtained at level 380. The birch pollen grains and one hickory might be taken to indicate pluvial climatic conditions. On the other hand, the major features of the spectrum are quite similar to those of strata immediately above. I doubt that the pollen is primary; possibly part or all of it infiltrated down from higher levels. The pink caliche is generally void of pollen.

Double Adobe III, Chiricahua Stage, Fig. 23. To represent a site containing artifacts of Chiricahua age, Mr. Sayles suggested a locality about 2.1 km. north of Double Adobe and 1 km. west of the Elfrida cutoff. At this point, the pink eroded surface of the valley fill has been entrenched by the present arroyo floor and is overlain by several meters of flood plain sediment. Samples were collected from a young (13th century?) channel fill and from older silts and blue clay immediately overlying the "pink caliche." The blue clay excavated by Sayles and Antevs had yielded Chiricahua type artifacts.

High percentages of grass and composites occur throughout the profile. No other pollen sections yielded as high a frequency of grass pollen. Changes in pollen composition elsewhere seldom involve a change in abundance of grass. The high value at Double Adobe III probably reflects local abundance of sacaton (*Sporobolus*). The abundance of composites suggests *Ambrosia* or its relatives.

From the surface to a depth of 140 cm. chenoams, grass, and composites are all relatively abundant, with little stratigraphic change. Below 140 cm. the abundance of grass pollen, the presence of sedge, and the nature of sediment indicate very wet conditions

locally. Sayles and Antevs characterized it as a "charco" which they believed was filled by sediment after a period of cutting during the altithermal. A radiocarbon date of 4,960 ± 300 agrees with the archaeological interpretation and makes unlikely the correlation of unit 1 with prealtithermal beds as presented in Martin, et al (123). The radiocarbon date accompanies a maximum in pine, juniper, sedge, *Typha,* and grass —suggesting a wet altithermal.

Malpais Site, Chihuahua. Fig. 24. Fossil bones of late Pleistocene *Equus* and *Bison* about 25 km. southeast of Nuevo Casas Grandes were encountered by Dr. Charles DiPeso of the Amerind Foundation. The fossil locality is a shallow arroyo about 3 meters in depth, roughly 2 km. west of the settlement of Malpais, elevation roughly 1,460 m. No artifacts were discovered. With the help of Dr. DiPeso I was able to collect the pollen section shown in Fig. 24 and in Photograph 14.

From the surface to 65 cm., cheno-ams are abundant, averaging 70 percent. Pine pollen is present but infrequent. Single grains of elm and hickory appeared at level 55 cm. Levels 75, 85, and 95 proved sterile. From level 105 to the bottom of the section there is a slight but significant decrease in cheno-ams. *Tidestromia* increases as do pine and oak. Twenty percent pine pollen was found in the deepest level in gravel. One elm pollen grain appeared in level 145. Presence of the extinct fauna and the pollen grains of hickory and elm, plus the abundance of cheno-am pollen suggests a correlation with pollen unit 2 at Double Adobe II and pollen unit 1 at Double Adobe I. Admittedly the distance between these localities is great and the possibility of independent ecological control in the two areas cannot be ignored.

The lack of any major change in pollen composition above and below the level with extinct animal bones is noteworthy. It is clear that a major climatic change did not occur. If the extinct fauna were able to occupy the area under early postglacial conditions, as seems certain, there is no clear-cut palynological evidence to account for its failure to do so in the late Pleistocene.

Murray Springs, Fig. 25. This locality, 2.7 km. west of Lewis Springs, is an arroyo tributary of the San Pedro River, elevation 1,243 m. It was visited in the hope of obtaining more information on the mammoth remains, radiocarbon sample, and artifacts reported by Sayles at Arizona EE:8:13 (181). Schoenwetter and I were unable to locate either the mammoth or Cochise Culture remains. We collected a pollen sequence in an arroyo bank close to or at the spot described as " . . . heavily carbonaceous earth, about 500 feet north-east of an old ranch house at the forks of drainage." Sayles' radiocarbon sample (A-69 8,250 ± 200 B.P.) came from about a meter below the surface of the ground.

From 0 to 60 cm. the pollen profile is dominated

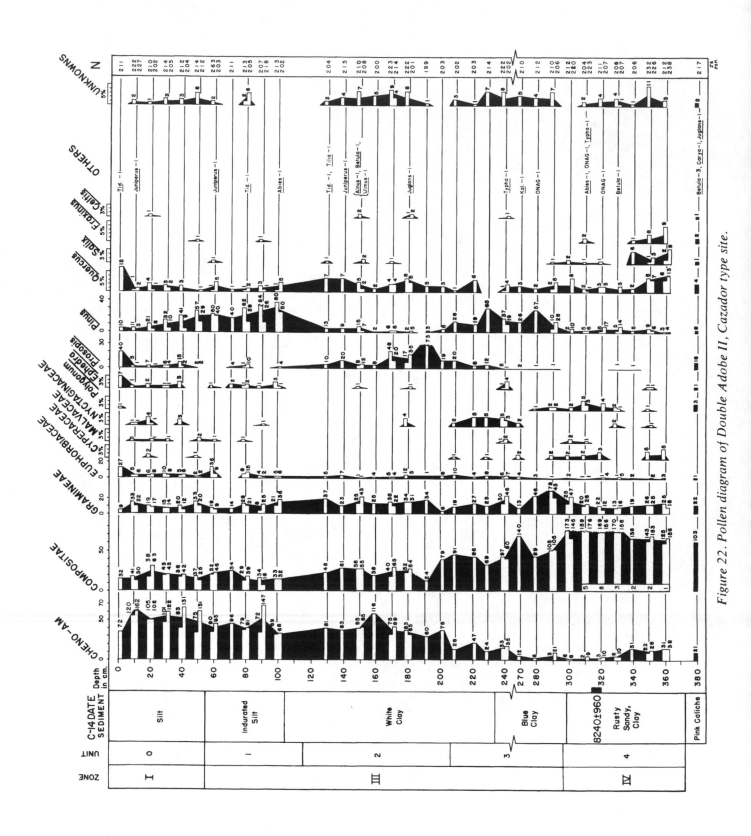

Figure 22. Pollen diagram of Double Adobe II, Cazador type site.

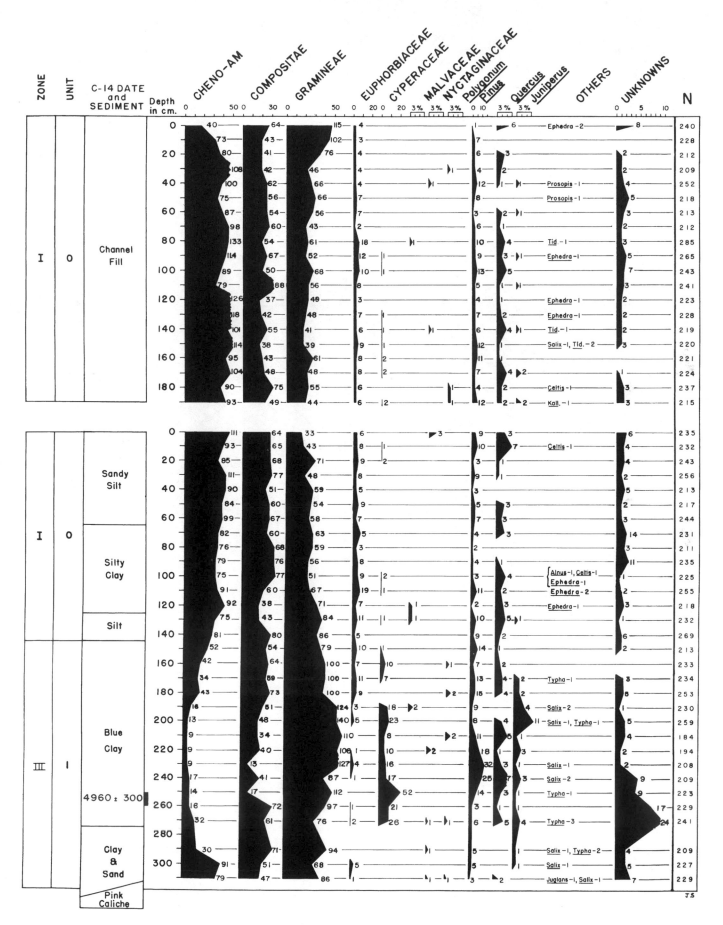

Figure 23. Pollen diagram of Double Adobe III, Chiricahua stage.

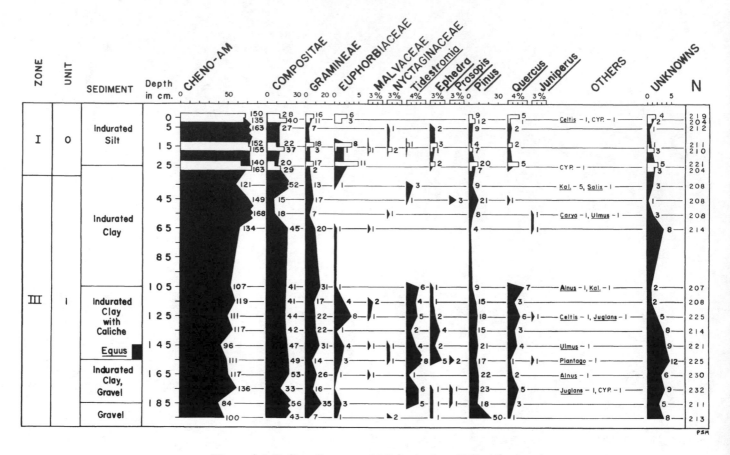

Figure 24. Pollen diagram of Malpais site, Chihuahua.

by cheno-ams accompanied by slightly more *Ephedra* than is found in pollen zone I elsewhere. Below 60 cm. to a depth of 100 cm. Compositae increases, cheno-ams decline somewhat, and *Ephedra* virtually disappears.

On the assumption that the radiocarbon date of 8,250 ± 200 collected by Sayles came from beds close to the Murray Springs profile and at a depth of about three feet, I sought to correlate pollen unit 1 at Murray Springs with pollen zone IV of the Double Adobe sequence. This required postulating that the lower portion of the Murray Springs diagram was older than pollen zone IV, and of late pluvial age (123).

Two C-14 samples collected at the time of pollen sampling, and recently dated, present the possibility of a more satisfactory correlation. The younger of the new dates, 4,150 ± 500, is slightly below the maximum of Compositae, which thus appears to represent pollen zone II, the younger of the two Compositae peaks seen in the postglacial sequence from Arizona, 4,000 to 800 B. P.

Pollen unit 2 should, according to the C-14 dates, represent the classic altithermal of Antevs. It is characterized by relatively high frequencies of cheno-ams and grass pollen with a peak in the summer rain-dependent Nyctaginaceae and with more pollen of sedge and cat-tail *(Typha)* than appear elsewhere in the profile.

Pollen unit 3, also part of the altithermal of Antevs according to the C-14 date, has more Compositae, including a maximum of Liguliflorae and a maximum of tree pollen.

The relative importance of pollen of hygric plants plus the relative abundance of pine and other tree pollen in units 2 and 3 is typical of the sites which can be assigned to the altithermal on the basis of C-14 dating. The tree pollen maximum is the basis for my inference that the altithermal was not a period of drought but rather of increased precipitation.

My failure to relate the Murray Springs profile to earlier surveys at this site underscores another principle of palynological studies in the Southwest: It is essential to coordinate pollen sampling with archaeological field work as closely as possible. In the present case, despite the best efforts of E. W. Sayles to help me relocate the original site, it appears that the Murray Springs profile cannot be related with confidence to the elephant bones, or to the Cochise cultural material, or to the original radiocarbon sample from this locality.

San Simon Cienega, Fig. 26. Throughout the Southwest a few cienegas (uneroded and undrained wet meadows) remain, the remnants of a flood plain environment typical of the early 19th century. The San Simon Cienega lies east of the Chiricahua Mountains just inside New Mexico, 4 km. north of the Cienega Ranch, T. 26 S., R. 22 W., sec. 1, elevation 1,180 m. It is approximately 8 km. long and 1 km. wide. It is dominated by sacaton

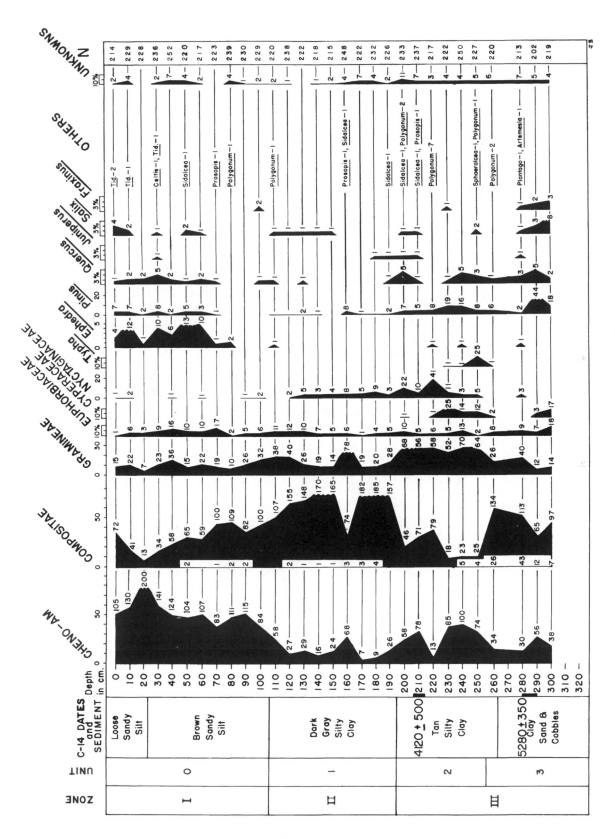

Figure 25. Pollen diagram of Murray Springs.

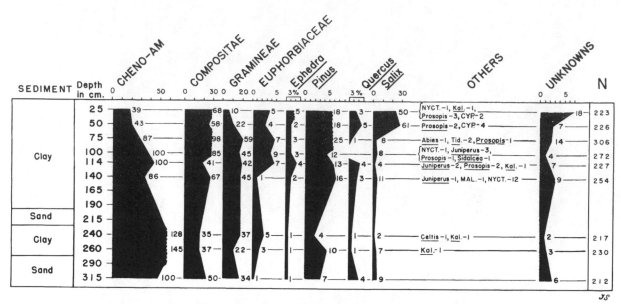

Figure 26. Pollen diagram of San Simon Cienega.

grass (*Sporobolus airoides*) with various Cyperaceae, other grasses, and a few willows. Creosote bush *(Larrea)* occupies enclosing bajadas. Two cement spillways built in the 1930's have prevented headward erosion by the San Simon Creek from cutting into the cienega and lowering the water table, which lies within a meter of the ground surface.

In August of 1956 Dr. E. B. Kurtz and I collected pollen samples with a Davis peat borer. Sand strata and indurated mud made collecting difficult. The resulting pollen diagram (Fig. 26) is unlike other pollen diagrams from the desert grassland. Compositae are relatively common near the top and cheno-ams increase with depth, the reverse of what one finds in dissected flood plains. Abundant composites, especially *Ambrosia,* grow today in the Southwestern cienegas and it is likely that the high frequency of Compositae pollen near the top of the San Simon diagram reflects a history of cienega conditions.

Lehner Site, Fig. 27. The best known association of Early Man and late Pleistocene extinct animals found in an alluvial deposit in southern Arizona is that excavated on Lehner Ranch, 2 km. south of Hereford (80). Initial efforts at extracting pollen from alluvial soil samples collected by R. Shutler at the Lehner site were unsuccessful (106). By repeated hydrofluoric acid treatments, P. J. Mehringer was able to recover sufficient pollen in all of Shutler's samples to prepare a pollen profile, characterized by dominance of Compositae in the lower portion, cheno-ams in the upper portion, and with a rise in *Ephedra* at the very top. Only the lower portion of the Compositae-dominated pollen unit is shown in Mehringer's diagram (Fig. 27).

The zone of Compositae dominance (mainly low-spine types), with Liguliflorae, represents both the black "k" horizon and the "1" horizon of Haury, et. al. CO_2 gas counting of the "k" bed resulted in a radiocarbon date of 10,410 ± 190 (A-33). It immedi-

Figure 27. Pollen diagram of Lehner site.

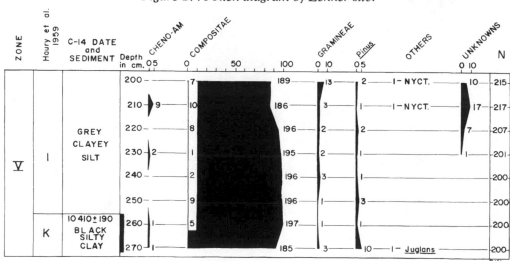

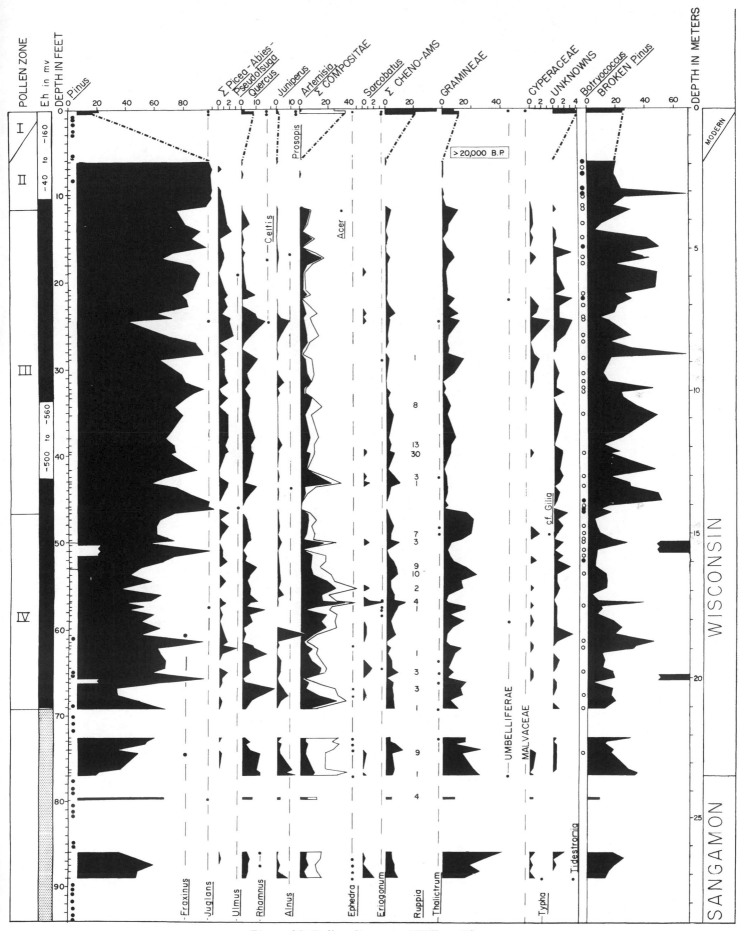

Figure 28. Pollen diagram of Willcox Playa.

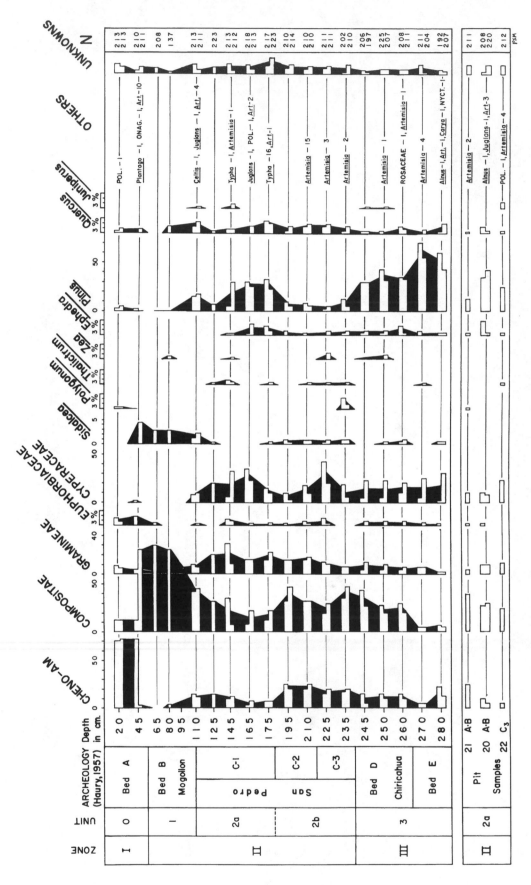

Figure 29. Pollen diagram of Cienega Creek site, Point of Pines.

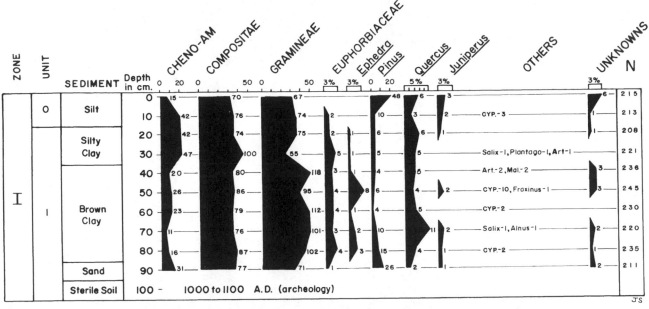

Figure 30. Pollen diagram of Dry Prong Reservoir, Point of Pines.

ately overlies bones of mammoth, horse, tapir, and bison associated with Clovis-type points in a gravel bed radiocarbon dated at 11,240 ± 190 (A-42, 42). Antevs (7) interpreted the black "k" bed as representing the deposit of a cat-tail and tule swamp and he correlated it with the last pluvial.

Results of pollen analysis (Fig. 27) indicate that the "k" and "1" beds contain no sedge (tule) or cat-tail pollen. The preponderance of low-spine Compositae closely resembles pollen unit 3 of Double Adobe I and the "k" and "1" beds at the Lehner site are readily correlated with pollen zone IV of the desert grassland chronology. The environment of deposition may have resembled the Empire Wash Cienega of the Empire Ranch, where high Compositae counts can be recovered from surface soil samples collected under colonies of giant ragweed *(Ambrosia trifida)*. At present the main conclusion to be drawn from the pollen record at the Lehner site is that it represents an environment similar to that recorded at Double Adobe in the initial phase of post-pluvial alluviation. There is no pollen evidence to suggest that the climate 10,000 years ago was appreciably different from the present climate of southern Arizona. Pollen evidence of *major* climatic change is found only in much older sediments, such as that of the Willcox Playa.

Willcox Playa, Fig. 28. Although my main purpose is to illustrate the pollen record of postpluvial time, it is helpful to compare the postpluvial results with a much older pollen profile. A 42 m. core collected from the middle of the Willcox Playa at 1,260 m. elevation in Cochise County provides a long pollen stratigraphic record and indicates the nature of the Southwestern climate through a major portion of the Pleistocene.

Without examining the results in detail, one can see that the upper portion of the Willcox core, C-14

dated at 23,000 years, differs strikingly in pollen composition from any of the desert grassland spectra of postpluvial age and also from the surface soil pollen spectra shown at the very top of the Willcox profile. The pollen record between 2 and 21 meters is dominated by pine, with small but regular amounts of *Picea, Abies,* and *Pseudotsuga* (abbreviated as PAP on the diagram).

Below three meters *Artemisia* and *Sarcobatus* appear in a higher frequency than they are known in the post-pluvial pollen record. The Wisconsin-age maximum of pine and other conifer pollen between two and three meters probably represents an interval when yellow pine parkland invaded the region now occupied by desert grassland (84). Only in the lower half of the core, between 36 and 42 meters, are there pollen spectra which can be compared with spectra from some of the flood plain pollen profiles.

Unfortunately, the pollen sequence of the Willcox Playa core is not continuous. The distribution of black dots at the left edge of the pine curve indicates samples which were extracted and found to contain insufficient pollen for a routine 200-grain pollen count. From the surface of the playa down to two meters is a zone in which pollen has been oxidized or otherwise destroyed. The buried pollen record of the Willcox Playa is much older than the oldest of the alluvial flood plain samples known from the desert grassland. It cannot be correlated with the flood plain pollen record of the desert grassland.

Pollen Profiles from the Yellow Pine Parkland

Postpluvial pollen stratigraphy need not be confined to the desert grassland environment. Two pollen profiles from below the Mogollon Rim in the vicinity of the Point of Pines Field School in the San Carlos Apache Indian Reservation represent pollen records from within the yellow pine (*Pinus ponderosa*) parkland. For facilitating field work of J. Schoenwetter and for giving generously of their special knowledge of this area I am indebted to Profs. E. W. Haury and R. M. Thompson.

Cienega Creek Site, Point of Pines, Fig. 29. Haury (76) has reported on stratigraphic and cultural features of a small flood plain of the Cienega Creek, Graham County. The locality lies 6.4 km. south of the Point of Pines Field School of the University of Arizona, elevation 1,880 m. It should not be confused with Cienega Creek of the Empire Valley in southern Pima County. Pollen samples from the Cienega Creek site near Point of Pines were collected by Haury during excavation and can be related to depositional beds described at the site (76).

Martin and Schoenwetter (121) reported a total of 42 corn pollen grains throughout the profile. This was based on a scan of 57,000 pollen grains; *Zea* found in the 200-grain count only is shown on Fig. 29. The archaeological record, extending from San Pedro or possibly Chiricahua stages of the Cochise sequence into the Mogollon, makes this the longest reasonably continuous record of *Zea* pollen encountered in an alluvial deposit from the Southwest.

The pollen diagram presents several features that distinguish it from other Southwestern profiles. Pine declines from a maximum value of 50 percent to less than one percent. Sedge virtually disappears near the surface. Above 125 cm. *Sidalcea* (a mallow rare in the profiles of southern Arizona) is common, unusually so for a zoophilous species. The spectacular shift from Compositae to cheno-ams, which appeared to occur within two sub-samples of the 45 cm. level, may represent relatively recent cutting and lowering of the water table. In the upper five levels pollen preservation was very poor, with insufficient pollen present in level 95 cm. for a 200-grain count.

It is clear that the archaeological beds recognized by Haury at the time he collected the pollen samples coincide quite closely with major pollen units. On the other hand a correlation of the Cienega Creek site profile with the pollen zones found in the desert grassland of southern Arizona is less obvious. The zonal correlation shown at the extreme left of Fig. 29 is based on archaeological age estimates. Despite discrepancy in results of radiocarbon dating from the Cienega Creek site, discussed by Haury (76) and Damon and Long (42) it appears that unit 3 should not exceed 5,000 years in age; it may be much younger. The high frequency of pine suggests a correlation with pollen zone III of the desert grassland.

Regarding climatic change, it is tempting to mount a case for regional climatic dessication in terms of diminishing pine pollen frequency. An equally plausible interpretation is that as the flood plain aggraded, the alluvial surface suitable for growth of non-arboreal plants increased. Sedge, grass, and composite pollen production increased, diluting the regional upland forest pine pollen rain, as it does the modern samples from stock tanks and cienega soils in the meadow surrounding the Cienega Creek site (see Table 1).

A single unmistakable pollen grain of hickory (*Carya*) appeared in bed E, level 280 cm. Another was discovered while scanning for *Zea* in level 195. The records at Point of Pines are decidedly younger than most of those from southern Arizona. Did hickory linger in the pine clad uplands below the Mogollon Rim after it disappeared at lower elevations? Or are all records of *Carya* in Arizona to be explained as long distance transport?

Dry Prong Reservoir, Point of Pines, Fig. 30. Adjacent to a kiva complex excavated near Point of Pines by A. Olsen lay a depression, apparently manmade and used as a reservoir during occupation of the site. The site dates from the Reserve Phase of the Mogollon, so the base of the pollen profile from the reservoir must begin about 1000 to 1100 A. D. The location is roughly 40 km. northeast of the University of Arizona Field School, elevation 1,800 m.

A single major change in pollen content is evident. Above level 20 cm. the frequency of pine rises and Euphorbiaceae and *Ephedra* decline. The low but regular percentages of *Ephedra* through unit one are of interest. This shrub generally shuns yellow pine forest; it is a frequent feature of pinyon-juniper woodland. In an unpublished inventory of 200 species in the Point of Pines flora, V. L. Bohrer omits mention of *Ephedra*. None was seen by Martin or Schoenwetter during field work in the area.

In a careful scan of modern tank and soil surface samples (N = 22,235) Schoenwetter found 25 *Ephedra* grains (0.1 percent). The pollen record plainly indicates a greater abundance in prehistoric time.

The post-occupation drop in *Ephedra* (levels 0, 10 cm.) and the increase in pine pollen frequency appear to occur at about the time when agricultural activity ended and the region was abandoned, around 1450 A. D., according to Haury (77). An increase in tree growth, with a decline in heliophytes as *Ephedra*, could be explained on cultural grounds. It could also signal a change to a colder, wetter climate with improved conditions for tree growth. The fact that no contemporaneous increase in pine pollen was found in alluvium of the desert grassland may favor a local cultural explanation for increasing pine pollen frequency and the decline in *Ephedra* in the post-occupation history of the Dry Prong site.

VI. COMMENTS ON THE POLLEN TYPES

Most of the pollen found in postpluvial sediments from the arid Southwest can be assigned to three plant families; Compositae, Gramineae, and Chenopodiaceae (including *Amaranthus*). Although the nearest pines may be 25 to 50 km. away, pine pollen invariably occurs in alluvial deposits from the desert grassland. Animal-pollinated plants are poorly represented in alluvium but the following groups occur regularly: Nyctaginaceae, Malvaceae, *Tidestromia, Kallstroemia, Euphorbia,* and *Eriogonum.*

Virtually all pollen grains found in postpluvial alluvium represent pollen of plants known to occur in Arizona in historic times. The exceptions are *Tilia, Ulmus,* and *Carya.* Their occurrence could represent long-distance transport, but an early postpluvial invasion, at least of *Ulmus* and *Carya,* appears more likely (see discussion below).

Cheno-ams (Chenopodiaceae plus Amaranthus). The genera *Atriplex, Chenopodium,* and *Amaranthus* probably account for most of the periporate, psilate pollen counted as "cheno-ams." *Suaeda* is morphologically similar but prefers more alkaline soils than were found near most of the fossil localities. *Erotia* is a shrub of certain upper bajadas in the desert grassland, a habitat far enough removed from the mid-valley flood plains not to contribute much pollen to them.

Sarcobatus, another member of the *Chenopodiaceae,* can be identified by its relatively small number of annulated pores, superficially suggesting *Plantago.* It was not recognized until the later phases of the study and some *Plantago* determinations may be misidentified *Sarcobatus.*

Cheno-am pollen predominates in most of the modern pollen samples collected in southern Arizona (Fig. 7). The ecology of cheno-ams is distinctive. They prefer fine alkaline soils of flood plains and disturbed ground. They tend to avoid exposed bed rock or coarse well-drained soils of the upper bajadas. I believe high frequencies of cheno-ams in flood plain sediments, such as occur in pollen zones I and III, represent episodes of channel cutting with lower water tables in the flood plain and an accompanying rise in alkali content of the alluvial soil. When no deep channels occur and the water table rises, the less alkaline soils may be occupied by Compositae (especially *Ambrosia* and *Franseria*), Cyperaceae, and Gramineae.

In sediments from Lake Patzcuaro, Mexico, Hutchinson, Patrick, and Deevey (93) recognized two rises in cheno-am frequency which they thought might reflect prehistoric cultivation of grain amaranths. In southern Arizona the periods of cheno-am predominance cannot be directly related to prehistoric cultivation. The cheno-ams increase and Compositae decline at a time when prehistoric farming was on the wane in the Southwest.

Compositae. This family numbers 151 genera in Arizona alone. The subfamily Liguliflorae and the genus *Artemisia* can be recognized by their pollen; the wind-pollinated species generally have much shorter spines than the animal-pollinated composites. The importance of separating Compositae into high-spine (over three microns) and low-spine (under three microns) groups was not recognized until the end of the study. The low-spine group includes *Ambrosia, Iva, Franseria, Xanthium,* and *Hymenoclea.*

The times of Compositae dominance (pollen zones II and IV) represent intervals when low-spine composite pollen occurs almost to the exclusion of other pollen types. On ecological grounds it is unlikely that *Hymenoclea* (which is partial to sandy washes and avoids fine alluvium) and *Iva* (scarce in southern Arizona) have contributed much pollen to alluvial deposits. I have found 70 to 80 percent low-spine Compositae in surface soil samples collected beneath or near colonies of giant ragweed, *Ambrosia trifida* (Fig. 13, Table 3). *Ambrosia* may be main source of the Compositae which dominate pollen zones II and IV; other likely sources include *Xanthium* and annual species of *Franseria.*

Zea and other Gramineae. The frequency of grass pollen is relatively constant in the profiles of Cienega Creek and Matty Wash, Empire Valley, in Double Adobe I and II, in the Malpais Site, the San Simon Cienega, the Cienega Creek site, and the Dry Prong Reservoir. Grass pollen shows a maximum frequency in those pollen profiles which fall within the altithermal, i.e the lower parts of the pollen diagrams from Double Adobe III, IV, and Murray Springs. At Double Adobe III and at Murray Springs the grass pollen maximum accompanies relatively high frequencies of sedge pollen, suggesting a wet meadow.

Beyond the possibility of a grass pollen maximum during pollen zone III (the altithermal), the pollen count of grass does not disclose an appreciable change through the last 10,000 years. In deeper parts of the core from the Willcox Playa the frequency of grass increases, approaching the highest values of grass pollen found in postglacial alluvium.

Of the Arizona grasses only *Zea* (corn) can be identified — on the basis of its large size (over 60 microns). The pollen record of corn is irregular (Table 5). Although one might expect to find corn in all sites where pottery is present, it was absent or very scarce in most flood plain alluvium associated with the ceramic period. An exception is the 42 grains found in alluvial samples from the Cienega Creek site, Point of Pines (121). One level of the Cienega Creek profile in the Empire Valley contained corn pollen almost exclusively, an anomalous occurrence. The rest of the profile yielded only seven pollen grains of *Zea.*

Despite the apparent scarcity, very low frequencies of *Zea* in alluvium may be sufficient to indicate corn cultivation in the immediate vicinity. An irrigated field of hybrid corn near Kansas Settlement contained only 1.4 percent *Zea* in surface soil samples. *Zea* is found in high frequency (over 10 percent) in prehistoric refuse at Wetherill Mesa, Mesa Verde National Park.

At both the Point of Pines Cienega Creek site and in the Empire Valley corn pollen was found in pre-ceramic-age deposits of Cochise cultural affinity. The Empire Valley records may extend in time to slightly over 2,000 B. P. and the Cienega Creek site to at least 2,500 B. P. according to C-14 dates and probably much older on the basis of archaeology (76). The Bat Cave record of *Zea* in New Mexico 5,600 years ago would indicate that the climate of the "altithermal" must have been more suitable for corn cultivation than the climate of the last 800 years, when corn cultivation declined.

Euphorbiaceae. Tricolporate, finely reticulate grains with a transverse furrow are assigned to this family; perhaps all such pollen represents the genus *Euphorbia*. It is the most numerous of the zoophilous plants in the flood plain pollen record, occasionally reaching ten percent of the total count. No significant stratigraphic changes are evident in the pollen profiles.

Cyperaceae. Sedge pollen may be considered diagnostic of marshy ground, especially if accompanied by pollen of *Typha*. The Cienega Creek site, Point of Pines, contained more sedge pollen than any other flood plain under study; the preceramic period of occupation must have been a time when the flood plain at the Cienega Creek site was much wetter than subsequently.

Malvaceae. Most mallow pollen in the flood plains is a large, triporate, annulated type with very long spines, superficially resembling a large "high-spine" Compositae pollen grain. *Sphaeralcea* and its relatives *Abutilon, Gayoides,* and *Horsfordia* are the most likely pollen sources. The *Hibiscus* and *Gossypium* types are quite different and were not found. *Sidalcea* is also distinctive — a periporate pollen grain with spiral pores. The spines may be lost in fossilization. *Sidalcea* reached a maximum of four percent near the top of the Cienega Creek site profile.

Nyctaginaceae. Large, periporate, echinate and heavy-walled pollen with an unusually thick endexine is typical of the spiderling, *Boerhaavia,* a common annual growing after summer rains in the desert grassland. Other genera with similar pollen which may have contributed to the record of Nyctaginaceae include *Mirabilis, Oxybaphus,* and *Allionia. Abronia* and its close relative *Tripterocalyx* are tricolpate, thus morphologically distinctive. None of the remaining five genera known in Arizona are likely to have occurred close to the study areas. *Boerhaavia*-type pollen is especially numerous in altithermal-age beds, as at Murray Springs, suggesting heavy summer rains and disturbed soils.

Polygonum. All grains of *Polygonum* belonged to the section Persicaria, heavily reticulate and periporate pollen with pores occupying only a few of the lumina. On geographic grounds *P. coccineum* is the species most likely to be represented. It inhabits ditches, ponds, and marshes. *Polygonum* pollen was especially numerous near the bottom of Double Adobe I and II.

Another member of the Polygonaceae, *Eriogonum,* is a very common genus with many species in Arizona. Its large, prolate, tricolporate grain with distinctive wall

Table 4. Frequency of **Ephedra** types in the Southwest.

	Location	Age estimate	torreyana type	nevadensis type	Percent of torreyana type
	1. Southern Arizona stock tanks	Modern	17	1	94.4
	2. Mexico, Chih., Malpais Site	Post-glacial	20	0	100.0
	3. Arizona, Double Adobe I	"	19	1	95.0
	4. Arizona, Double Adobe II	"	20	0	100.0
	5. Arizona, Double Adobe III	"	12	0	100.0
torreyana dominance	6. Arizona, Double Adobe IV	"	7	1	87.5
	7. Murray Springs	"	64	0	100.0
	8. San Simon Cienega	"	18	3	85.7
	9. Cienega Creek, Empire Valley	Late post-glacial	9	0	100.0
	Southern Arizona desert grassland	Post-glacial	149	5	96.8
	10. Snaketown Canal, Chandler, Ariz.	900-1900 A.D.	220	8	96.5
	11. New Mexico, Dark Canyon	Post-glacial	184	0	100.0
	12. New Mexico, Fort Sumner	1100-1900 A.D.	38	0	100.0
	13. New Mexico, Chaco Canyon	1100-1900 A.D.	25	7	78.1
	14. Arizona, Point of Pines	Late post-glacial	4	44	8.3
	15. Ariz., Vernon; N.M., Reserve	Modern	21	32	39.6
	16. Arizona, Glen Canyon	Late Post-glacial	5	19	20.8
nevadensis dominance	17. Arizona, Navajo Canyon	"	12	76	13.6
	18. Arizona, Rampart Cave dung	Early post-glacial	0	100	0.0
	19. Utah, Escalante River	Late post-glacial	9	31	22.5
	20. Colorado, Mesa Verde Site 1200	Pueblo II-III	0	71	0.0
	21. Colorado, Mesa Verde Site 1205	Late post-glacial	6	42	12.5
	22. California, Panamint Valley	Modern	0	107	0.0

structure (conspicuous columellae) was recognized too late in the project to be removed from the unknowns.

Ephedra. Mexican tea (*E. trifurca*) is an uncommon shrub in the desert grassland in southern Arizona. It is more abundant along old beach lines north of the Guzman sink in northern Chihuahua and on gypsum soils of the Jornada Range, New Mexico (32). *Ephedra* is one of few anemophilous shrubs in the desert grassland, a feature that accounts for its importance in the fossil record. Significant changes in the postglacial pollen rain of *Ephedra* were encountered at Murray Springs and at the Point of Pines Dry Prong site.

Wodehouse (182) and Andersen (2) recognize two types of *Ephedra* pollen grains. The *torreyana* type has a large number of straight furrows and ridges and no or very faint hyaline lines intersecting the furrows; the *nevadensis* type has a smaller number of furrows; the ridges tend to be wavy, with conspicuous hyaline lines intersecting the furrows.

Steeves and Barghoorn (171) have examined 43 of the currently recognized 48 species of *Ephedra* and divided them into four types. The divisions are arbitrary, with continuous variation in the following characters which they studied: number of furrows and structure of ektexine ridges, size and form of intervening furrows, and nature of the hyaline lines, which they term "colpi." Of the Southwestern species it appears that their type D (*trifurca*) plus *E. antisyphilitica* and *E. torreyana* would be included within the *torreyana* group; their type A (*E. clokeyi, coryi, funera, viridis*) plus *californica, nevadensis,* and *aspera* lie within the *nevadensis* group.

In the analysis of profiles from various parts of the Southwest, I have encountered a sharp geographic segregation in the *torreyana* and *nevadensis* groups (Table 4, Fig. 31). In southern Arizona, New Mexico, and northern Mexico the *torreyana* type occurs with a frequency of 90 to 100 percent. Hafsten (75) found it predominant in west Texas. In northern Arizona, southern Utah, and southern Colorado there is a sharp change and the *nevadensis* type dominates (Table 4). Point of Pines in central Arizona has less of the *torreyana* type than one might expect.

The general pattern is fairly clear cut; the *torreyana* type occupies the Mexican border region and the *nevadensis* type is concentrated from the Four Corners area of northern Arizona and southern Colorado, westward into the Great Basin and the Mojave Desert (Fig. 31). The *torreyana*-type species of the Mexican Plateau and its environs occupies a region of summer precipitation. The *nevadensis* type prevails in a region of win-

Figure 31. Centers of Ephedra *pollen types in the Southwest.*

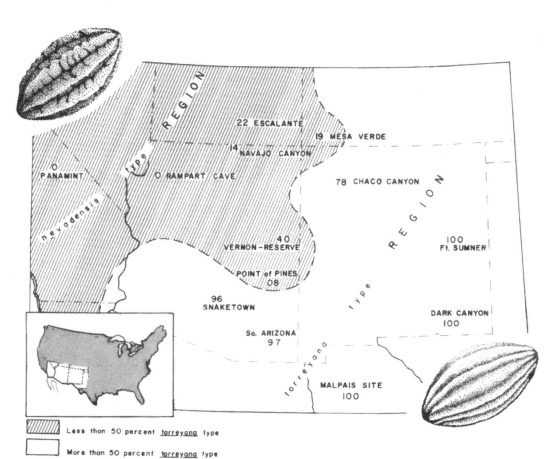

ter precipitation. Overlap in *torreyana* and *nevadensis* types occurs in parts of the Southwest where the two rainfall components meet and overlap.

If the morphological segregation is consistent and if it is related to seasonal distribution of rainfall, it may be possible to determine qualitatively the past history of winter and summer rainfall. In this regard, I note that there is no apparent change in frequency of *torreyana* from early to late postglacial sediments in southern Arizona. If the type is climatically significant, it means that southern Arizona has experienced summer anti-cyclonic precipitation throughout the postglacial period.

Artemisia. *Artemisia* pollen reaches values of 10 to 20 percent in the Willcox Playa core (Fig. 28). Six grains were found in the clay ball sample at Double Adobe I (Fig. 20) and the more numerous records of *Artemisia* in the early postglacial suggests lingering survival of pluvial age populations. *Artemisia filifolia,* the only shrubby member of the genus presently found in southern Arizona, occupies sandy ridges east of the town of Willcox (131). Pollen from the Willcox population did not appear in a stock tank sample 12 km. west southwest of the sand ridges. Herbaceous species of *Artemisia* must account for most of the postglacial records.

The virtual absence of shrubby *Artemisia* from the Southwestern deserts and desert grassland and the relatively minor importance of herbaceous *Artemisia* within the desert grassland at present means that it may prove a good indicator of glacial-age climatic change in the Southwest. At present *Artemisia* contributes a very small amount of pollen to stock tanks and other modern traps of the desert grassland. During the Wisconsin and earlier glacial periods it was much more common in the Southwestern pollen rain.

Prosopis. Range-management ecologists have commented repeatedly on the brush invasion of ranges from Texas to Utah in the past 50 years (89). Mesquite is perhaps the foremost invader; others include *Juniperus, Larrea, Haplopappus,* and *Gutierrezia.*

Most of the shrub invaders either lack a distinctive pollen morphology or shed very small amounts of pollen. An exception is mesquite which is an important minor component in many pollen diagrams. The pollen grains are tricolporate with long colpi, subprolate and scabrate. Under conditions of good preservation a delicate and distinctive oval annulus may be seen around the pores.

The recent shrub invasion of upland ranges is attributed basically to reduction in range fire frequency, enabling mesquite and other shrubs to compete more effectively with the native grasses and forbs (89). The presence of mesquite in the postpluvial pollen record is not in conflict with evidence for recent mesquite expansion into Southwestern ranges. The likely source for fossil mesquite pollen would be riparian trees of the flood plains where it has a long history; it is doubtful that the spread of mesquite into the desert grassland

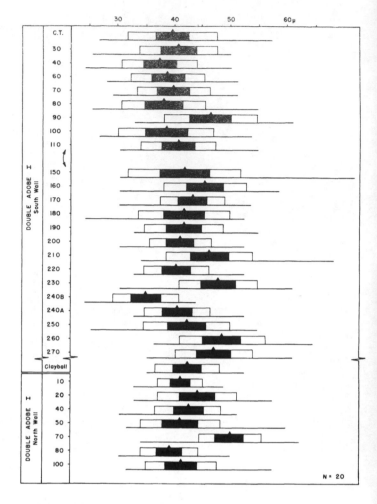

Figure 32. Pine size-frequency at Double Adobe I, Sulphur Spring stage.

would affect pollen content of the flood plain alluvium. There is a slight increase in mesquite pollen frequency at the top of 3 out of 5 profiles in which it occurs regularly, perhaps marking a very recent increase of mesquite along the flood plains.

Pinus. Kearney and Peebles (96) list nine species native to Arizona. Past changes in frequency of pine are a reasonable index of climatic change. A high frequency of pine pollen should accompany periods of increased moisture or lowered evaporation; low frequency should accompany periods of maximum warmth or dryness. An additional method of providing information on past conditions is size-frequency analysis.

A relatively rapid and efficient method of plotting mean bladder length is the graphical size and comparison of Hubbs and Hubbs (87). The mean, range, twice the standard error of the mean, and one standard deviation on either side of mean are plotted for each level — not two standard deviations and one standard error as stated incorrectly by Martin (117).

Fig. 32 of Double Adobe I shows significant differences in means between a number of adjacent levels. However, there is no obvious trend from the early to the late postpluvial.

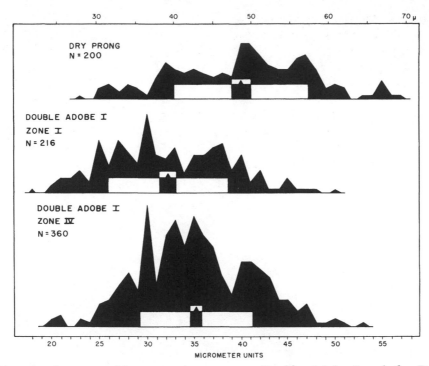

Figure 33. Pine size-frequency histograms for strata at Double Adobe I and the Dry Prong site.

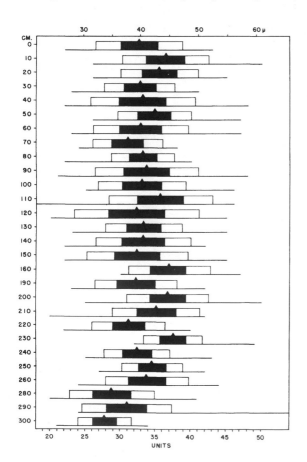

Figure 34. Pine size-freqeuncy at Murray Springs.

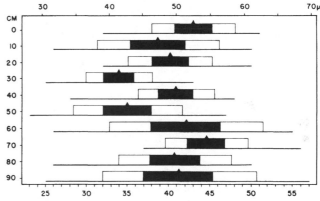

Figure 35. Pine size-frequency at Dry Prong site, Point of Pines.

Measurements of the early and the late postpluvial interval were compared with those from a yellow pine-pinyon region, the Dry Prong site at Point of Pines (see size-frequency histogram, Fig. 33). Means of each population are significantly different. While pollen of the early postpluvial pines from Double Adobe is larger than that of the late postpluvial and may contain less pinyon pollen, both are smaller in mean size than the Dry Prong series from yellow pine parkland.

Size statistics of pine pollen from Murray Springs reveal fewer significant shifts between adjacent levels than appeared at Double Adobe I. A possible trend may be found in the pollen record at the bottom of the profile (Fig. 34) where there is a significant shift to small-size populations, implying a relatively greater representation of pinyon. This occurs in levels 290-300 which have a higher frequency of pine pollen than other parts of the profile.

Measurements at Dry Prong (Fig. 35) reveal high variance and significant shifts in means between certain adjacent levels. There is no clear-cut trend as one might expect if pinyon had been replaced by yellow pine after abandonment. The means of the Dry Prong populations are higher than pine pollen means from Murray Springs and Double Adobe I.

Size statistics will not reveal which species of pine occupied Arizona in the last 10,000 years. But size statistics should be of considerable value if the changes which appear in certain horizons can be used as stratigraphic markers. Cause of the changes (levels 80, 90, 100; levels 220, 230, 240B, 240A; levels 50, 70, 80 in Double Adobe I, Fig. 28) is unknown. Two possibilities may be considered: first, a short-term change in pollen production differentially favoring either pinyons or non-pinyons; secondly, that some conditions of preservation were temporarily altered and differential shrinkage or expansion of all pollen occurred in that level. In core samples from the Valley of Mexico Sears, *et al.* (158) found considerable size variability due to differences in preservation and chemical treatment.

To conclude the paleoecological interpretation of pine pollen, it can be shown that relative pollen percents are a function of elevation and distance to the nearest pine trees. Pine pollen size, a parameter independent of the pollen count, should also be useful in revealing stratigraphic changes and may ultimately suggest something about species composition. Major change in mean pine pollen size which was found between glacial and postglacial periods in eastern Pennsylvania (117) are not seen in the postglacial pine pollen record of Arizona.

Fraxinus-Salix. Both of these trees grow along suitable flood plain environments, usually close to permanent water. *Salix* and *Fraxinus* are still found in the Whitewater Draw drainage at the point where Leslie Creek flows through the Swisshelm Mountains and permanent water reaches the surface. They are absent at Double Adobe. High values of these genera in early postglacial time indicate a riparian community and surface ground water along the Whitewater flood plain. A high water table can be assumed after storage of runoff during the pluvial period ending about 10,000 B. P.

Carya-Ulmus-Tilia. In a total of 142,000 pollen grains counted or scanned from postglacial age pollen sections seven pollen grains of *Carya*, five of *Ulmus*, and two of *Tilia* were discovered (Table 5). An additional 26,000 pollen grains in sloth dung at Rampart Cave (122) yielded one grain of *Carya*. *Ulmus* appeared in Pliocene sediments near Prescott (73) and in early Pleistocene sediments at Safford (74). Pollen of each of these genera is unmistakable and misidentification is most unlikely.

Although *Ulmus* and *Carya* are often cultivated, neither *Carya, Ulmus,* nor *Tilia* is part of the native flora of New Mexico, Arizona, or northwestern Mexico. Considering their very rare occurrence as fossils (Table 5) one might imagine that the pollen of these trees represents long-distance transport into the Southwest from a distant eastern pollen source. *Carya* and *Ulmus* range west almost to the Big Bend of western Texas.

Table 5. Occurrence of rare or noteworthy pollen types in postglacial-age sediments of southern Arizona and northern Chihuahua.

Locality	Grains counted & scanned	Abies	Picea	Acer	Alnus	Betula	Carya	Juglans	Ulmus	Tilia	Fraxinus	Artemisia	Zea
Cienega Creek, Empire Valley	6,390												7+
Matty Wash Empire Valley	8,226					1		3				7	3
Double Adobe IV	8,590	1			2			1			1	2	1
Double Adobe I North wall	7,510				1		1	9			48	3	1
Double Adobe I South wall	7,200	1			2	1	2	3			16	31	1
Double Adobe II	16,000	2		1	4	7	1	8	1	1		7	
Double Adobe III	14,770		1		1			2					
Malpais Site, Chih.	4,150				2		1	2	2				
Murray Springs	8,100				1						14	1	
San Simon Cienega	2,170	1											1
Cienega Creek Site, Point of Pines	56,940				3	2	2	7	2	1			42
Dry Prong Site, Point of Pines	2,440				1			1			1	2	1
TOTAL	142,500	5	1	1	17	11	7	36	5	2	80	53	57+
Modern samples southern Arizona	7,660	6	2		1			15			2	18	

The hypothesis of long-distance transport deserves critical appraisal. Compared with the total number of grains scanned and counted, the frequency of *Carya, Ulmus,* and *Tilia* is indeed very low. Against this fact must be placed the observation that the distinctive pollen grains of certain other wind-pollinated trees native to parts of the Southwest are also quite scarce in both the ancient and the modern pollen rain. *Alnus,* a riparian tree or shrub of mountain streams, which grows in the Catalina and Pinaleño Mountains, was represented by 17 pollen grains in the sum of all postglacial counts; *Picea* by one; *Abies* by five; *Juglans* by 36.

In scanning slides for *Carya* and *Ulmus* care was taken to record all *Juglans*. That walnut, a close relative of *Carya* and a frequent native tree along the larger flood plains of the Southwest, should be represented by only 36 pollen grains in the *entire* postglacial record suggests that low frequency alone is no proof of scarcity, even in an anemophilous genus. Hickory pollen is one-fifth as numerous as walnut. If the pollen of walnut is derived from trees growing along Arizona flood plains in postglacial time, why not that of hickory?

A second very important point is the absence of any hickory or basswood from modern soil or stock tank samples (N = 30,000 including scan) and from the late postpluvial, that is pollen zones I and II. Elm occasionally appears in stock tanks, an occurrence to be attributed to pollen of the Chinese elm commonly grown as an ornamental in towns and on ranches. If long-distance transport brought *Carya* into Arizona in the early postpluvial period, why not in the last 4,000 years? If the source of these genera were upwind, the case for long-distance transport might be more convincing. In the Rio Grande Valley of southern New Mexico sugarbeet pollen was recovered at an elevation of 1,500 m. above the valley floor (129). Prevailing winds might carry atmospheric pollen hundreds of miles from such an elevation. In the case of elm, hickory, and basswood, the source area is to the east of Arizona. Except during late June, July, and August, weeks after the flowering period for these trees, the prevailing winds blow from the west, and long-distance drift would be far more likely to occur from Arizona and New Mexico into Texas, rather than in the opposite direction.

A third point supporting the case for *Carya* in the early postglacial is the record of hickory wood from Double Adobe, identified by Prof. L. H. Daugherty (151). In recent correspondence Daugherty (letter of January 29, 1960) states that he is unable to locate the Double Adobe specimen. He reports that the wood of *Carya* is distinctive and that there is no reason to question the determination.

Supporting the case for postglacial *Ulmus* west of its present limit is the discovery of six pollen grains in early postglacial-age cave fill from Dark Canyon Cave, near Carlsbad, New Mexico (unpublished study by J. Schoenwetter). The pollen is associated with a late Pleistocene megafauna. The profile indicates no major change in climate, and the presence of elm is the only aberrant feature of the pollen record. Both *Ulmus* and *Carya* were found in various Pleistocene sediments from New Mexico (75).

The matter of *Tilia* is more of a puzzle. *Tilia* has no Pleistocene megafossil record in the Southwest. Presently it occurs much farther east than *Carya* and *Ulmus*. Only two *Tilia* pollen grains have been encountered to date. Postglacial invasion of this region by *Tilia* seems highly unlikely. Perhaps the *Tilia* grains represent redeposition of Tertiary material.

If the postglacial occurrence of *Ulmus* and *Carya* can be accepted, what is their paleoecological meaning? Antevs inferred pluvial conditions from the presence of *Carya* at Double Adobe. But hickory pollen did not appear in analysis of pluvial age deposits from the Willcox Playa. Elm was found only twice in the playa core. During pluvial times hickory and elm may not have occupied the Southwest. I suggest that a specialized riparian environment suitable for hickory and elm existed only in the early postpluvial period. With intense channel cutting of the hypsithermal, which removed most of the early postpluvial flood plain sediment, the edaphic conditions suitable for elm and hickory also were destroyed. Such an event would not reflect direct climatic control, although climatic change (presumably intense summer rainfall) may have initiated hypsithermal cutting.

Abies, Picea, and Alnus. In cool wet forests of the highest desert mountains occur small stands of *Abies* (Chiricahuas, Huachucas, Santa Ritas, Santa Catalinas, Pinaleños) and *Picea* (Chiricahuas and Pinaleños only). *Alnus* occupies wet ravines of the Santa Catalinas, Rincons, and Pinaleños. None of these trees occur on the bajada slopes or along arroyos of the valley flood plains. Occasionally their pollen appears in postglacial alluvium or stock tank sediment 10 to 100 km. from the mountains where these trees occur. Such erratic occurrence can be expected through southeastern Arizona. The combined pollen frequency of *Abies* and *Picea* in all postpluvial counts (N = 142,000) is 0.01 percent. The record of one to two percent *Abies* and *Picea* from the Willcox Playa demonstrates a past increase in frequency of these trees.

Betula. Birch does not presently occur south of the Mogollon Rim of central Arizona. Pollen of birch might have blown from this point to the Cienega Creek site, Point of Pines, where two grains were found. On the other hand the presence of seven birch pollen grains in southern Arizona is less likely to represent wind transport from northern Arizona.

Betula was rare in the pluvial sediments of the Willcox Playa (Fig. 28), the top of the pink caliche below the postpluvial pollen section at Double Adobe II, and in the early postpluvial deposits at Double Adobe (see Table 5). One grain was found in a mud chip containing selenite from a well cutting of the valley fill at a depth of 150 m., 2 km. south-southeast of Double Adobe at the Owensby Ranch. With it were 17 cheno-ams, 2 grass, 3 composites, and 1 *Celtis* pollen grain. This could represent an early Pleistocene pollen record. Gray (74) found two grains of *Betula* and 24 of the palynologically similar *Ostrya-Carpinus* type in early Pleistocene cores near Safford.

The record of *Betula* from the top of the pink caliche at Double Adobe II is probably secondary. Attempts at extracting pollen from the highly calichified surface of valley fill have been generally unsuccessful.

Despite three birch and a hickory grain at 380 cm. depth at Double Adobe II, the very close correspondence in all other features between this and the early postglacial pollen deposit at level 360 immediately above cannot be ignored. Most of the pollen in the pink caliche, perhaps all of it, may be contamination from the younger beds.

If birch invaded southern Arizona in the glacial period, as seems probable, the early postpluvial records at Double Adobe in Sulphur Spring stage sediments may have come from relict colonies lingering in the Chiricahuas and Pinaleños. The relict colonies did not survive into late postglacial time.

VII. POLLEN STRATIGRAPHY AND RADIOCARBON DATING

For purposes of correlation, I have relied largely on radiocarbon dates (Tables 6 and 7), especially those associated with the pollen profiles. Archaeological and paleontological age criteria were used to a lesser degree. Radiocarbon dating of channel walls presents some special problems. The ideal source of organic carbon, visible pieces of charcoal, is not always present where one wishes. Finely divided, alkali-insoluble organic material composed of lignins, charcoal, pollen and spores, and other non-humic fragments is disseminated through most alluvial deposits, especially those dark in color, which typify buried cienega soils.

This finely divided carbonaceous material may be contaminated by both older and younger carbon. The older carbon could include rebedded particles. In such case there should be evidence of redeposition in the pollen record, evidence which I have either not found or not recognized. Furthermore, such a large amount of organic debris on the soil surface is moved into channels by floods that it is unlikely that finely disseminated carbon rebedded from older channel fill would be a major source of organic material in the headwater flood plains under study.

Younger carbon may be intruded into older beds

Table 6. Radiocarbon dates associated indirectly with pollen profiles. Cochise County, Arizona, except as noted (42) (43).

Sample no.	Locality	C-14 age determination	Pollen profile and unit dated	Comments
A-33	Lehner mammoth site	10,410±190	Composite maximum	Layer "k" of Haury, et. al.
A-42	" " "	11,240±190	Below composite maximum	Bone bed
A-67	Double Adobe, Sulphur Spring site, FF:10:1	9,350±160	Double Adobe I, unit 3	Charred wood in river sand associated with artifacts of the Sulphur Spring stage
C-216	" " "	7,756±370	" " "	Charcoal-bearing dirt, Sulphur Spring stage
A-69	West of Lewis Springs, Ariz. EE:8:13	8,270±260	Murray Springs, relation to pollen units uncertain	Sample was collected at a depth of 3 ft.
A-28	Point of Pines, Graham Co., W:10:112	1,900±160	Cienega Creek site, Bed C-2	Carbonaceous material
A-20	" " "	2,100±150	" "	Charcoal
A-23	" " "		" " , Bed C-3	Wet, charred wood
A-51	" " "	3,190±160	" " " "	Rotten, wet wood
A-26	" " "	2,490±170	" " " "	Organic material
A-26B	" " "	2,900±150	" " " "	" "
A-27	" " "	3,190±150	" " , Bed D	Charred, water-logged pine branch
A-25A	" " "	2,700±160	" " " "	
A-25B	" " "	2,440±160	" " " "	
A-29	" " "	2,430±150	" " " "	Scattered charcoal fragments

by the growth and eventual decay of roots. Mesquite is a notorious inhabitant of flood plains, especially dissected valleys, where its roots readily penetrate five to 10 meters of alluvium to reach the water table. A living root — likely of mesquite — was found 52 m. below ground surface at Mission Mine near Tucson (42).

Despite careful efforts at removing roots, finely divided and only partly humified root fragments must contribute to the "carbonaceous material" commonly dated in arroyo samples. Separate measurements of the radioactivity of the macroscopic charcoal fraction and the associated carbonaceous particles were made in three cases (samples A-184, 188, and 189). None of the paired results exceed laboratory error, although the difference between fractions of A-184 is quite close to being significant, and in each case the carbonaceous alluvium fraction is the younger. The "inverted ages"

seen in samples A-189 and A-191 and in A-193 and A-194 each represent a date on charcoal above a younger date on carbonaceous alluvium. The former are likely to be more accurate age estimates.

The humic fraction, often removed in sample preparation, was counted in sample A-227. The result is a younger, but not significantly younger age than that of the carbonaceous fraction, indicating persistence of ancient humic acids in buried flood plain soils.

Organic and inorganic fractions of sample A-192 differed by 2,600 years in their C-14 ages. While the carbonaceous material may be "too young," it is also possible that the inorganic fraction incorporated ancient $CaCO_3$, and is accordingly "too old." Ancient carbonate is less likely to be incorporated in lake deposits but the possibility remains; both A-362 and A-353 are dates on inorganic carbonate and both may exceed

Table 7. Radiocarbon dates directly associated with pollen profiles from Cochise County, Arizona, except as noted (42) (43).

Sample No.	Pollen Profile	Presumed Archaeological Affinity	C-14 age	Dominant Pollen Type	Depth and Unit Dated	Comments
1. A-352	Willcox Playa	none	23,200±500	pine	6'3" to 7'0"	Inorganic carbonate
2. A-353	" "	"	22,000±500	"	7'8" to 8'5"	" "
3. A-188c	Double Adobe I	Sulphur Spring stage	8,270±250	composites	unit 3	Charcoal
A-188e	" "	" "	8,260±160	"	" "	Carbonaceous alluvium
4. A-189	" "	" "	8,960±100	"	" "	Charcoal
	" "	" "	8,680±100	"	" "	Disseminated charcoal and organic material
5. A-191	" "	" "	8,000± 60	"	" "	Carbonaceous alluvium
6. A-190	" "	" "	7,910±200	composites, cheno-ams	unit 2	" "
7. A-184c	Double Adobe II	Cazador stage	8,240±960	composites	320 cm. below surface, unit 3	Charcoal
A-184e	" "	" "	7,030±260	"	" "	HCl leached carbonaceous alluvium minus hand-picked charcoal
8. A-192a	Double Adobe III	Chiricahua stage	7,560±260	grass	250-260 cm. below surface, unit 1	Inorganic carbonate
A-192b	" "	" "	4,960±300	"	" "	Carbonaceous alluvium
9. A-186	Murray Springs	none	4,120±500	grass, cheno-ams	210 cm. unit 3	Organic alluvium
10. A-187	" "	"	5,280±350	composites	270 cm. unit 4	Alluvium with low organic content
11. A-193	Double Adobe IV	San Pedro or Chiricahua stage	3,860±200	"	100-110 cm. unit 1	Soil sample with charcoal from floor of pit
12. A-194	" "	none	2,860±440	cheno-ams	235-245 cm. unit 2	Organic alluvium
13. A-88	Matty Wash, Pima Co., upper "peat" stratum	San Pedro stage	2,010±150	composites	Matty Canyon, unit 2	Plant fragments buried in alluvium
14. A-92	Matty Wash, Pima Co., lower "peat" stratum	" "	2,220±150	"	" "	" "
15. A-227A	Cienega Creek, Pima Co.	" "	2,140± 60	"	630-650 cm. below top of arroyo	Carbonaceous alluvium
16. A-227B	" "	" "	1,790±400	"	" "	Humic material (NaOH soluble fraction of the sample)

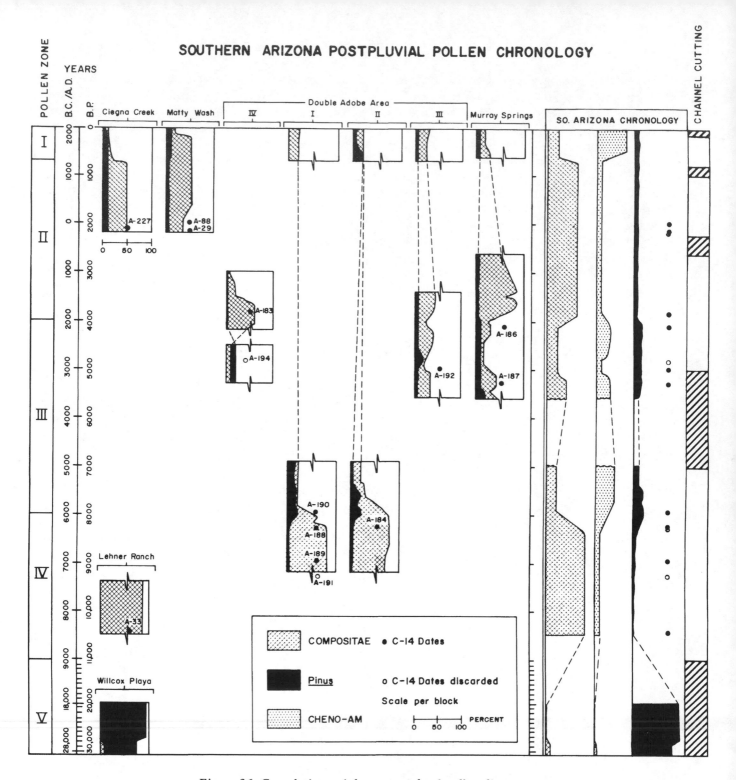

Figure 36. Correlations of desert grassland pollen diagrams.

the true age of the Willcox Playa sediments between 1.9 and 2.4 m.

By relying on radiocarbon dates and other stratigraphic information the pollen profiles from alluvium of the desert grassland of southern Arizona were correlated to form a postpluvial chronology (Fig. 36). The chronology is based on changes in abundance of cheno-ams, composites, and pine. It reveals two episodes of cheno-am dominance. But despite exposed stratigraphy, archaeological and paleontological control, and 15 radiocarbon dates directly associated with the pollen profiles, the postpluvial chronology presented for South-

western flood plains in the desert grassland incorporates a measure of uncertainty. The uncertainty stems from the possibility that the same type of pollen dominant did not always prevail at the same time throughout the flood plains of the desert grassland.

The evidence in behalf of synchroneity of either cheno-ams or composites is based on (1) the observation that one type, the cheno-ams, are widespread at present and have been so for several hundred years; (2) the fossil record in separate valleys generally indicates either cheno-ams or composites dominant at the same time (Fig. 36); and (3) the pollen types used for correlation are thought to reflect either a high water table (composites) or a low water table (cheno-ams) in the flood plains. The flood plain water table should be closely controlled by stream regimen — the cutting or filling of the flood plain. As the geological evidence of synchronous cut-fill episodes throughout the Southwest is fairly convincing, it is also likely that the flood plain pollen record will be similar in adjacent drainages.

In at least one case the fossil record reveals lack of synchroneity. Composites persist in sediments deposited by Matty Wash after the 1200's at a time when sediments containing a high frequency of cheno-ams were deposited along Cienega Creek (Fig. 36). The abundance of Compositae near the top of the profile of San Simon cienega also departs from the expected condition of cheno-am dominance in the last few hundred years. These inconsistencies provide us with some insight regarding ecological preferences of the two dominant flood plain pollen types. From the modern pollen record in Empire Wash it appears that abundant low-spine composite pollen signifies a slowly aggrading or relatively stable fine alluvial surface. The high frequencies of cheno-ams and composites which occur within 100 meters of each other in Empire Cienega shows that within a single flood plain atmospheric pollen mixing does not obliterate local differences in plant community composition. In the case of Matty Wash channeling occurred later than in adjacent Cienega Creek, with low-spine Compositae (probably *Ambrosia*) persisting along the Matty Wash flood plain after cheno-am invasion had occurred one kilometer to the west on Cienega Creek.

From the interesting and very suggestive record of surface samples in the Empire Cienega (Table 3, Fig. 13) and from the fossil record of the Compositae it appears that flood plains formerly dominated by composite pollen in southern Arizona represented cienega environments. The record of cheno-ams and their ecology indicates that they occupy dissected or eroding flood plains with a more alkaline soil and lower water table than those favored by the Compositae.

Although their ecological preferences may help in understanding the stratigraphic record, it is clear that the cheno-am and composite pollen content of sediments will not by itself provide information on the regional climate. The postpluvial pollen record of southern Arizona indicates minor climatic changes at most, a record of change comparable to that in adjacent areas north of the Mogollon Rim in eastern Arizona and New Mexico, in the San Juan Mountains of Colorado, and in the Llano Estacado of Texas and New Mexico (see Fig. 37). Since pluvial time there is no evidence of major climatic change in any of these records; the pollen sequence farther east in Soefje Bog in east Texas is remarkably stable (72).

The desert grassland pollen zones shown on each diagram and summarized in Fig. 36 may be characterized as follows:

Zone I. During the past 500 to 1,000 years cheno-ams dominate; pine, grass, and composites are of low frequency. The pollen rain in the late prehistoric and early historic period is quite similar to that of the present day. An exception is found in the Empire Valley where composites remain abundant into the 19th century (153). Pollen zone I can be correlated with zone C3b of eastern North America (49).

Zone II. Between about 4,000 years ago and the 13th century A.D. the pollen record is dominated by composites. Pine, grass, and cheno-ams are uncommon. *Zea* pollen is more likely to occur in zone II than elsewhere in the record. Under conditions of ponding or cienega formation the relative abundance of grass and sedge pollen may be high. These pollen types appear to be under local rather than regional control and are not readily used in correlations. Zone II extends into the sub-boreal portion of the European-American chronology (49).

Zone III. A prevalence of cheno-ams and grass and a pine pollen frequency often in excess of 15 percent is diagnostic of zone III. Such a condition was found in the pollen record beneath a 3,800-year radiocarbon date from the floor of a Late Cochise feature at Double Adobe IV. Relatively high frequencies of pine pollen, usually accompanied with moderately high frequencies of cheno-ams, occur at Double Adobe III and at Murray Springs between 4,000 and 5,500 B. P. Parts of zone III are represented in calichified silts and clay at Double Adobe I and II where both *Ulmus* and *Carya* may occur. The zone is placed between 4,000 and 8,000 B. P., esentially the altithermal of Antevs.

Zone IV. Composites are dominant (70 to 90 percent), declining only near the transition to zone III, where pine may become relatively abundant. Liguliflorae, *Fraxinus, Salix,* and *Artemisia* are generally more common in zone IV than elsewhere. Radiocarbon dates cluster between 8,000 and 9,000 B. P. (Table 6), exceeding 10,000 B. P. at the Lehner site. The Lehner diagram falls within the younger Dryas (park-tundra) interval in the Danish chronology, but no major climatic change is evident in the Lehner record.

Zone V. This is the Wisconsin pluvial. Pine is dominant with spruce, fir, and Douglas fir more abundant than in postpluvial time; cheno-ams and Compositae never in high frequencies. During drier intervals oak, *Artemisia,* and NAP increase in abundance. Exact thickness of Wisconsin pluvial deposits in the Willcox Playa is unknown. Between two meters and 40 meters there is a general decline in pine and a rise in NAP, especially grass.

No alluvium of pluvial age has been found in southern Arizona. Only at the Tule Spring site in southern Nevada is a pluvial (Wisconsin) age pollen record known in eroding flood plain deposits. The four postpluvial pollen zones presently proposed for the desert grassland were orginally presented as six by Martin et al. (123).

VIII. A POSTPLUVIAL CLIMATIC MODEL FOR THE ARID SOUTHWEST

From the pollen record and other fossil information, summarized in Fig. 37, it is possible to appraise the history of the desert grassland. The change in pollen content most useful as a bioclimatic index is the frequency of pine. In the last pluvial period, about 20,000 years ago, pine pollen reaches a very high frequency (over 90 percent) in the region new occupied by desert grassland. The pollen rain of *Artemisia,* spruce, fir, and Douglas fir was also heavier than it is at present. The regional vegetation was probably a yellow pine forest or coniferous savanna growing under the type of climate suitable for western yellow pine. The entire pollen assemblage suggests a major increase in winter precipitation. Whether or not summer rains persisted is uncertain; the upper half of the Willcox Playa core included none of the summer rain elements present in postglacial time such as *Kallstroemia, Tidestromia,* and Nyctaginaceae.

Between 20,000 and 12,000 years ago the pollen record is broken in southern Arizona. Judging from the scarcity of pine pollen, the oldest sediments in the flood plain pollen sequence represent a dry period. Pollen zone IV at both the Lehner site and the Cazador and Sulphur Spring type sites, near Double Adobe, has very little pine or other tree pollen. On the other hand, an abundance of ash and willow characterizes pollen zone IV at Double Adobe and suggests a high water table. Ash and willow formed a narrow ribbon of gallery forest along the Whitewater Draw, as they do presently at the edge of the Swisshelm Mountains along Leslie Creek. Good ground water conditions throughout the Southwest may be expected in early postpluvial time when storage in the valley fill was at a maximum following pluvial recharge. The ash-willow record represents riparian conditions along the Whitewater Draw; the scarcity of upland pine and oak pollen lends no support to the interpretation that the Sulphur Spring stage was moist and is of late pluvial age (151, 27, 6). 10,000 years ago desert grassland had returned to southern Arizona.

In sediments overlying those containing Sulphur Spring stage artifacts, and contemporaneous with a late Pleistocene megafaunal horizon, the frequency of pine pollen rises slightly (Double Adobe I, II), signifying more moisture. Presence of a maximum of *Polygonum* and sedges indicates wet riparian conditions perhaps with tules *(Scirpus)* along the Whitewater Draw. In Unit 2 of Double Adobe II pine pollen frequency is low again and mesquite is at a maximum. The decline in pine is less apparent in Double Adobe I and the mesquite invasion is evident only in levels 150, 160, and 170. At the Malpais site pine is slightly but significantly more abundant near the bottom than near the top of the profile. The record of oak is generally rather uniform in all sites, with no major changes that might convey stratigraphic or paleoclimatic meaning.

Hickory and elm are present in very small amounts and essentially disappear above zone III. The presence of hickory and elm does not reflect major climatic changes of pluvial age. There is no resemblance in pollen content between the beds in which hickory and elm occur and the pluvial age lake mud examined from the Willcox Playa.

Proceeding to the events which immediately postdate the extinction of the late Pleistocene megafauna, we encounter a crucial period in postglacial time. Flood water erosion gouged broad arroyos in the inner valley alluvium. Thick caliche deposits formed in the freshly cut arroyo channels and walls. Antevs (5) dates the period he calls altithermal at 7,500 to 4,000 B. P. Was the Southwestern climate especially dry then?

At least three pollen diagrams represent a portion of the altithermal, as follows:

(1) At Double Adobe IV two meters of sediment extend below the floor of a Late Cochise (San Pedro stage?) feature. The excavated face was indurated with small caliche nodules but was not chalky white and heavily indurated as are the beds near Double Adobe. No conspicuous erosion surface was encountered. In its high cheno-am ratio and high pine count the pollen record can be compared with zone III of Double Adobe I and II. From the presence of mesquite and the high frequency of pine and oak a climatic interpretation of warm-moist conditions is indicated. The scattering of *Kallstroemia* and Nyctaginaceae, typically found in summer rain climates, indicates the major seasonal source of the precipitation. A radiocarbon date on charcoal above the pine maximum was 3,860 ± 200. The pine maximum should accordingly represent the altithermal.

(2) According to Sayles and Antevs (151) the Chiricahua stage of the Cochise Culture extended from the Sulphur Spring to San Pedro stages and thus encompasses all of or most of the altithermal of Antevs. Presumably a record of sediments contemporaneous with a deposit of Chiricahua-age artifacts would be most likely to yield a pollen record of altithermal age. With this hope in mind E. B. Sayles directed me to a locality in which he had found Chiricahua-type artifacts. These came from blue clay on the east bank of the Whitewater Draw, two km. north of Double Adobe. The resulting profile, Double Adobe III, shows little change and is remarkably high in grass pollen throughout, indicating enduring cienega conditions. For purposes of climatic interpretation it is evident that based on the pine, juniper and grass pollen record the regional climate was slightly more moist in strata radiocarbon dated at 4,960 B. P. Locally along the Whitewater Draw there was a very wet cienega or series of continually moist depressions with rank grass, sedges, and an indicator of ponded water, *Typha.* It is difficult to reconcile the

pollen record with a climatic interpretation of extreme drought.

(3) At the Murray Spring site C-14 dates of 4,120 ± 500 and 5,280 ± 350 bracket an interval when pine pollen is slightly more common than in other levels and when a maximum in sedge, grass, and *Polygonum* pollen indicates a wet flood plain.

Pollen samples of postaltithermal sediments from southern Arizona generally contain less tree pollen than those associated with the altithermal. At the Dry Prong site in central eastern Arizona a rise in pine pollen and a decline in *Ephedra,* which occurred in the last few hundred years, is not necessarily the result of climatic change. It is more likely the recovery of tree growth in a region formerly disturbed and cultivated by farming tribes.

An increase in tree pollen commonly reported in sediments from the four corners area (Arizona, New Mexico, Colorado, and Utah) has been interpreted as evidence of better climatic conditions for forest growth during the last few hundred years (125). It may also signify a decrease in local weed (NAP) pollen production, to be expected when a pueblo or other prehistoric village was abandoned. An apparent increase in tree pollen would accompany the decrease in herb pollen following abandonment and the end of disturbance by man, without benefit of forest expansion or climatic change.

In brief, the major features of postpluvial climatic history are threefold: an initial arid period (zone IV) climatically equivalent to the present and dating 8,000 – 10,500 B. P.; next, a less arid interval (zone III) with an intensified monsoon rainfall, the climate perhaps no warmer but slightly wetter than at present, and corresponding in time to the classic "altithermal," 4,000 to 8,000 B. P.; and finally an arid period (zones I and II) closely resembling present conditions but with at least one possible shift in the seasonal distribution of rainfall and lasting from 4,000 B. P. to the present. There is a strong resemblance to the classic threefold postglacial pollen sequence of von Post.

IX. A REAPPRAISAL OF POSTPLUVIAL CLIMATES

Recent pollen analysis of pluvial lake sediments sustains the view of most geologists that glacio-pluvial climates of arid America were decidedly colder and wetter than the present. It is within the postpluvial period that the desert grassland pollen chronology fails to support the climatic interpretations of certain previous workers. Foremost is lack of pollen evidence of pluvial conditions at the Sulphur Spring stage type locality (profile of Double Adobe I).

According to Antevs (151) "The bones of the extinct animals, the much greater moisture indicated by the hickory, the permanent river, and the lakelet show that . . . the beds . . . were deposited during the last Pluvial, the correlative of the last glaciation in western North America." In 1942, Bryan visited the Sulphur Spring valley and confirmed Antevs' interpretation that the alluvial beds of the Sulphur Spring stage were deposited at a time wetter than the present (27). But the pollen record of Sulphur Spring stage sediments bears no resemblance to the pollen of pluvial-age sediments of the Willcox Playa and makes a pluvial climate and age virtually impossible.

A second and perhaps more provocative discovery is the high frequency of moisture indicators including tree pollen in the middle of the postpluvial pollen record. This suggests a moist, not a dry altithermal. It led me to re-examine the Bryan-Antevs model of postpluvial climates which is commonplace in the literature of Southwestern prehistory. The discord does not involve the Bryan-Antevs alluvial chronology, based on reasonably secure field evidence of synchronous episodes of deposition and erosion in Southwestern flood plains. Since first clearly proposed by Bryan (25) the alluvial chronology has proved of great value to field workers. It has been extended from Arizona and New Mexico northward to Wyoming (105) and south to Central Mexico (26).

Regarding climatic interpretation of the alluvial chronology, Bryan, Antevs, and others, have incorporated the following lines of field evidence in support of a warm-dry altithermal: (1) episodes of erosion and alluviation. (2) megafaunal extinction, (3) caliche formation, (4) eolian deposits, (5) Pueblo abandonment, arroyo cutting, and the late 13th century record of narrow three rings, (6) contemporaneous geologic-climatic events elsewhere. To these we may add pertinent biogeographic evidence. The issue at stake is whether the geologic and biologic evidence leads to a valid climatic inference — does it provide conclusive proof of a mid-postpluvial drought? I will attempt to show otherwise.

Alluviation and Arroyo Cutting. The cause of arroyo cutting after an interval of alluviation is variously interpreted (reviews in 3, 62, 156). Events other than climatic change can initiate a switch from filling to cutting. Both increased and decreased precipitation have been adopted as explanations of arid land erosion, the former at least since the time of Dutton (54). More recently Quinn (146) and Garner (66) are among those who attribute cutting in dry regions to increase rather than decrease in rainfall.

Climatic conditions prevailing during alluviation should be recorded in the fossil record of the sediments.

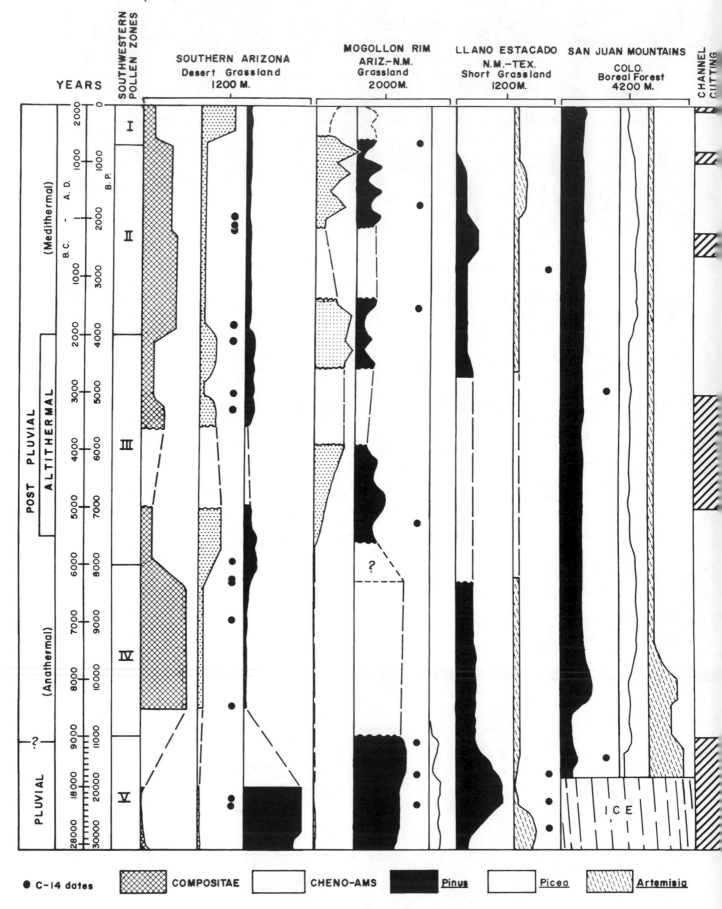

Figure 37. Correlation of the late Pleistocene pollen sequence in the Southwest.

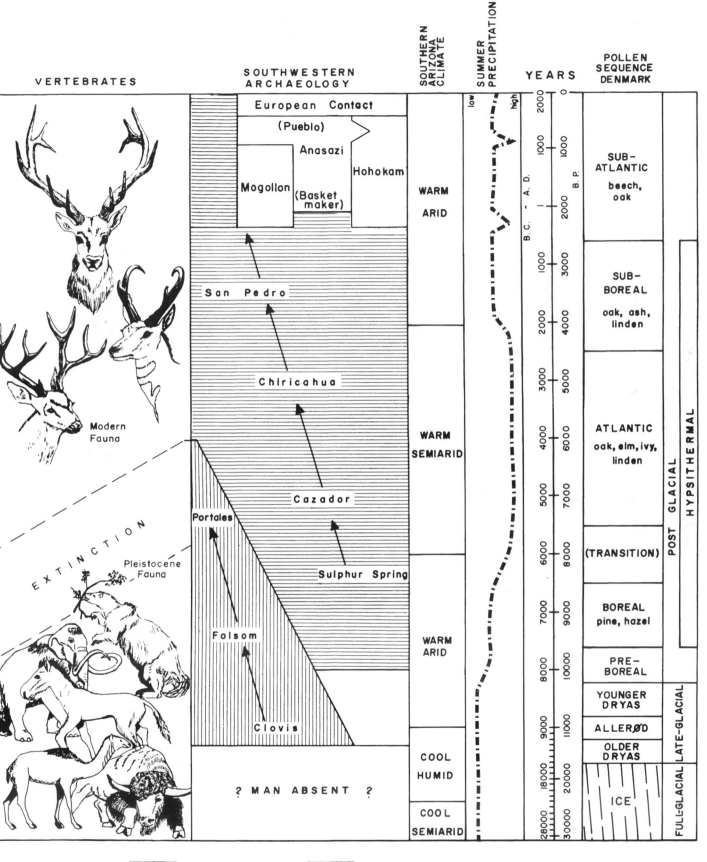

VERTEBRATES

Modern Fauna

EXTINCTION

Pleistocene Fauna

SOUTHWESTERN ARCHAEOLOGY

European Contact

(Pueblo)

Anasazi

Mogollon   (Basket maker)   Hohokam

San Pedro

Chiricahua

Cazador

Portales

Sulphur Spring

Folsom

Clovis

? MAN ABSENT ?

SOUTHERN ARIZONA CLIMATE

WARM ARID

WARM SEMIARID

WARM ARID

COOL HUMID

COOL SEMIARID

SUMMER PRECIPITATION

low   high

YEARS

B.C. – A.D.   B.P.

2000
1000
2000
1000   3000
2000   4000
3000   5000
4000   6000
5000   7000
6000   8000
7000   9000
8000   10000
9000   11000
18000   20000
28000   30000

POLLEN SEQUENCE DENMARK

SUB-ATLANTIC
beech, oak

SUB-BOREAL
oak, ash, linden

ATLANTIC
oak, elm, ivy, linden

(TRANSITION)

BOREAL
pine, hazel

PRE-BOREAL

YOUNGER DRYAS

ALLERØD

OLDER DRYAS

ICE

POST GLACIAL   HYPSITHERMAL

LATE-GLACIAL

FULL-GLACIAL

COCHISE CULTURE (Desert Tradition)   PALEO-INDIAN

Table 8. Alluvial chronology of Western United States (modified from Miller 1958) compared with pollen zones of the desert grassland.

FLOOD PLAIN EVENT	FEATURES	POLLEN ZONE	SUMMER RAINS
cutting 4	Began in late 19th century (Bryan 1925)	I cheno-ams	light to heavy
deposition 3	Began in most places A.D. 1200-1500[1]; continuing until 1880 or later.	I cheno-ams pine rise in northern Ariz., N.M.	" "
cutting 3	Coincided with decline of flood plain farming in many areas.	I cheno-ams	heavy
deposition 2	Two-fold in several areas; upper part at most places no younger than A.D. 1100-1200[1]. Lower part dates 2200-2400 B.P.[2], occasionally to 4000 B.P. Corn pollen rare but present throughout.	II composites	light
cutting 2	The altithermal of Antevs; calichification, dune formation along certain drainages. More extensive arroyo cutting than at any time since the last pluvial. First record of corn pollen.	III cheno-ams, pine	very heavy
deposition 1	A scattering of dates suggests an age of 7000-11000 B.P.[2]. Last record of the late Pleistocene megafauna.	IV composites	light
cutting 1	The major period of erosion in valley bottoms which defined the channels filled by alluvium in postpluvial time. Probably of Wisconsin glacial age and older.	V pine, other conifers in lake sediments.	very light to absent, winter precipitation at a maximum.

[1] Pottery and dendrochronology
[2] Radiocarbon (years before present).

Miller and Wendorf report (136): "Flora and fauna, along with such physical features as buried pond deposits, clearly indicate deposition under conditions more moist than at present." Although they considered such evidence to be known from many parts of the Southwest, they noted that direct fossil evidence for climatic fluctuation is exceedingly sparse in the alluvium they studied in the Tesuque Valley. At Chaco Canyon, Bryan (28) was careful to point out that the postpluvial alluvial deposits there did not produce incontestable fossil evidence of a wetter climate in the past. Regarding erosion and alluviation in the Near East near Jarmo, Wright (185) took the precaution to comment that his climatic interpretation of drought during the postglacial optimum rested largely on the occurrence of erosion at that time, not on fossil evidence of aridity. Other geologists have not always been equally candid in identifying their presuppositions.

If alluviation occurred under wetter climates and cutting occurred during or following drought, as is commonly assumed, one might expect to see such evidence in the pollen record. Such evidence is not apparent. On the contrary, pollen evidence of wetter conditions during the postpluvial period was found in sediments closely associated with the time of altithermal erosion.

The oldest period of alluviation, prior to altithermal erosion, is commonly considered to represent deposition under relatively humid conditions. This conclusion is based on presence of extinct animal remains and it raises the question of the ecological and climatic

meaning of the mammoth, camel, and other fossil mammals.

Extinct Megafauna. Fossil evidence often viewed as proof that deposition occurred under humid climates is the presence of extinct animal remains in certain early postpluvial alluvial beds (deposition 1 of Table 8). The suggestion that elephants, horses, camels, and other large mammals of western North America succumbed to drought following the last pluvial period can be traced at least to the time of Russell (149). Russell's view has become embedded in the scientific literature (111, 1, 27, 104, 91, 78, 79, 7, 62). An essential part of the theory is the notion that to thrive large animals require luxuriant vegetation. For example, regarding New Mexico near the San Jon site, Judson (95) wrote: "It is inconceivable that a mammoth could survive on the meager vegetation and surface water of today . . ." Is there biological evidence for such an opinion?

Prehistorians and geologists might recall the reflections of Darwin (46) "that large animals require a luxuriant vegetation has been a general assumption which has passed from one work to another; but I do not hesitate to say that it is completely false, and that it has vitiated the reasoning of geologists on some points of great interest in the ancient history of the world. The prejudice has probably been derived from India and the Indian islands, where troops of elephants, noble forests, and impenetrable jungles are associated together in everyone's mind. If, however, we refer to any work of travels through the southern parts of

Africa, we shall find allusions in almost every page either to the desert character of the country, or to the numbers of large animals inhabiting it."

Geologists and prehistorians might also recall that the African elephant (*Loxodonta*) survives along the southern border of the Sahara and that it ranged north to Fayum in the Libyan Desert in mid postglacial time (51). Even less likely victims of drought-imposed stress were the native American camels, llamas, and horses which also disappeared with the mammoth (118, 83, for radiocarbon dates and their evaluation). Evidence that one of the ground sloths (*Nothrotherium*) excelled as an arid lands herbivore, consuming aromatic or spiny shrubs which are shunned by domestic livestock, is found in the plant content of its feces (10, 122). Feral burros (*Equus*) occupy the most arid parts of the Mojave and Great Basin Deserts, a habitat their late Pleistocene predecessors abandoned 8,000 years ago. It is by no means obvious that in the absence of civilization large herbivores, including elephants, would not be able to thrive in arid America today. Nor is it obvious that climatic change played an important role in the extinction of large Pleistocene animals.

Pleistocene geologists have generally inferred major changes in the past climate of the Southwest, changes which can be clearly seen in the pollen content of ancient pluvial lake sediments. But the fact that the Pleistocene megafauna disappeared soon thereafter is no proof of a common cause. If climatic change, specifically drought, were responsible, why weren't small mammalian genera and other small vertebrates equally affected? Why didn't some plants disappear? How did relict populations of fishes manage to survive in small headwater springs and ponds of southern Arizona? Above all, what prevented the widely ranging mammoths, camels, horses, and ground sloths from enduring, if not in the Sonoran and Chihuhuan Deserts, at least in more mesic adjacent habitats such as the short grassland of the Texas High Plains, the mesquite grassland of the Mexican Plateau, or the tall grass prairie of the Midwest? Why should mastodons and ground sloths, which occupied both moist temperate and moist tropical habitats, have disappeared? Surely not because of "drought"!

This latter point was raised by Russell who recognized that the theory of aridity as a cause of extinction in the west could not be extended eastward. Russell's appraisal of the extinction problem includes another pertinent biological observation: "Even if it could be shown that a period of extreme aridity preceded the present climate, it is difficult to understand why certain mammals, as the elephant, mastodon, horse, camel, megalonyx, etc., should have become extinct, while others no more capable of withstanding great changes of environment should have survived" (149).

The biology and chronology of Pleistocene extinction invites additional study. But to believe that the entire Pleistocene megafauna of the Southwest could have been wiped out by drought is biologically naïve. Climatic inferences based on presence or absence of large mammal remains alone are best abandoned. I believe the fossil pollen record will prove to be a far more reliable guide to past vegetation and climate. It indicates past intervals of wetter as well as dry conditions in arid America. The Pleistocene megafauna survived all climatic change until the arrival of Early Man.

Caliche Formation. Antevs, Bryan, and others have observed that the old alluvial deposits (deposition 1) which were eroded and exposed during the altithermal are more heavily indurated than those of last postpluvial age. My limited field studies confirm this observation. Caliche nodules typify the older alluvial beds (deposition 1 of Table 8), and caliche is especially abundant in soil surfaces at the top of valley fill throughout southern Arizona.

Shreve (169) reported less caliche in the hot, low plains of Sonora than in the Arizona upland. Brown (20) claimed that caliche will not form in soils under either a completely arid or completely humid climate. Bretz and Horberg (19) follow Bryan and Albritton (29) in viewing successive caliche beds as a record of relatively arid conditions, with resolution occurring under relatively humid conditions. The basis for their conclusion rests largely on the geographic distribution of caliche and does not resolve the matter of its paleoclimatic significance in the regions where it occurs. Within a single paper one author altered his climatic interpretation of caliche from that of warm-dry to cool-dry (8).

Carbon-14 content of caliche in desert soils near Tucson does not support the concept that it is actively forming at present (Sigalove, pers. corr.). From its C-14 content it appears that much of the caliche in soils in southern Arizona may have formed in pluvial times. Until more is known about the formation of caliche paleoclimatic deductions based on its presence or absence are premature.

Eolian Deposits and Dune Formation. A well-known paradox of Southwestern deserts is the scarcity of large sand dunes. Where dunes occur the sand source usually lies immediately down wind of a large river as the Rio Grande or the Lower Colorado, or down wind of an ancient shore line as the upper Gulf of California, Guzman Playa, Tularosa Basin, Salton Sea, and Willcox Playa. The Hopi dunes of northern Arizona are unrelated to a major shore line and owe their origin to weathering of an extensive sandstone outcrop.

In west Texas, Huffington and Albritton (88) correlate an older dune deposit with the post-Neville disconformity, and a younger deposit with the post-Calamity disconformity in Albritton and Bryan's Trans-Pecos alluvial chronology. Although they infer aridity during dune formation there is no fossil evidence to support this claim, and the exact age and climatic meaning of the dune fields are uncertain.

It is commonly assumed that desert dunes will form when plant cover is weak, but in the absence of a source of sand no dunes are to be found, whatever the nature of plant cover. The major dune deposits of the Southwest appear inactive at present and they may have formed under glacial-pluvial conditions of the Pleistocene (132). Extreme aridity cannot be inferred from dune sand alone, independent of other evidence (62).

"The Great Drought." A crucial matter in the consideration of postglacial events is the well-known relationship between deep arroyo cutting (erosion 3 of Table 8), narrow tree rings, and Pueblo abandonment in the 13th century. The narrow tree rings found in ancient pine beams from A. D. 1215-1299 are taken as evidence of drought of a magnitude unknown in historic time (155). Such evidence has been accepted as conclusive proof that prehistoric erosion occurred under drought conditions (136).

At this point we may again consider the matter of summer vs. winter precipitation. To speak of drought in terms of *annual rainfall* is legitimate, providing one can demonstrate that the condition under study is sensitive to rainfall deficiency in both seasons.

The inutility of narrow tree rings as an index of *summer* rainfall is stressed in a neglected paper on tree rings and droughts by Gladwin (69). He points out that "both Douglass and Schulman have claimed that correlations can be established between tree-growth and *winter* precipitation, but each of them has been explicit in saying that *summer* rainfall has little, if any, effect upon the growth of trees" (p. 9). Gladwin was mainly concerned with the relationship between tree rings and Pueblo Indian crops, and deplored the fact that: "The concept that narrow tree-rings are indicative of drought conditions, which were capable of affecting the welfare of Pueblo peoples in ancient times, has become firmly fixed in the lore of Southwestern archaeology" (p. 32).

Gladwin's point is documented in the important table of correlations of ring growth and rainfall assembled by Schulman (155). Whereas the *annual* record for rings follows that of adjacent stations with a correlation coefficient usually above $+0.40$, a correlation between summer rains (July-September) and adjacent stations is very low or absent and may be positive or negative ($r = -0.20$ to $+0.23$, only four values above $r = +0.10$ in 23 trees from Mesa Verde, Colorado, south to Tucson, Arizona).

If "drought-sensitive" trees are to be the basis of prehistoric climatic interpretations, one wonders what they can tell us about *summer* rainfall. The most one might infer is that from 1276 to 1299 *winter* precipitation was abnormally low. Whether this would result in a drop in total precipitation and runoff is less certain. A compensating increase in summer rain could

cause arroyo cutting. If a drought sufficient to shrink upland vegetation zones occurred in the 13th century, I have failed to spot it in the late prehistoric pollen record from southern Arizona.

Biogeographic Commentary. A well-marked contingent of montane pine-oak woodland plants and animals evidently entered (or reentered) southern Arizona and New Mexico from Mexico rather recently. Some of the populations in question are isolated in the desert mountains (for example, *Sceloporus jarrovii, Crotalus lepidus,* various species of *Sonorella, Thomomys botteae*) and are unable to spread across desert and grassland barriers. To determine their rate of evolution one must establish the age of the ecological barrier isolating the montane populations. The difficulty is more serious than might be imagined; in the absence of a Pleistocene fossil record, unknown for many of the smaller animals, the evolutionist may not be certain whether he is dealing with isolation during the last 5 to 15 thousand years or during the last 5 to 15 million.

Palynological evidence of major environmental shifts in the late Pleistocene of the Southwest include (1) an interlude of juniper-pine woodland in the Mohave Desert (147), (2) a past forest of spruce and pine in the San Augustin Plains (35), (3) pine woodland or forest in the Texas High Plains (75) and southeastern Arizona (84), (4) Upper Sonoran conditions in the lower Grand Canyon and vertical displacement of vegetation zones 1,200 m. (122). Although pollen evidence was lacking at the time, Antevs (4) concluded that a 1,200 m. shift in life zones occurred in New Mexico in the last pluvial period. From the evidence of radiocarbon dates the last major pluvial maximum would be slightly older than 20,000 years, with pluvial conditions ending by 12,000 B. P. (Fig. 37).

The pollen record of pine and *Artemisia* with a low, regular frequency of *Abies, Picea,* and *Pseudotsuga* in the Willcox Playa (Fig. 28) may be sufficient to provide the record missing when Shreve wrote about the adjacent Pinaleño Mountains (163): "There seems to be no evidence, however, in favor of such a radical change of climate within the life of these mountains as to have brought about a continuous lowland vegetation like that which is now confined to the mountain summits." Admittedly, one or two percent of spruce and fir in sediments from the shore of the Willcox Playa (84) does not suggest spruce-fir forest of the character one finds on the top of the Pinaleños at 3,000 m. However, the change in pine should signify a major vertical displacement of spruce and fir as well, enough to greatly enhance their opportunity to spread across the intermontane divides, across what is presently desert grassland. We need not adopt, as Shreve did, a hypothesis of accidental arrival. Specifically, the evolutionist can view any population differences between montane pines, oaks, mammals, and land snails as likely to be the result of *not more than 20,000 years of isolation.*

The minor pine rise evident in pollen zone III points toward a second opportunity for immigration of montane elements. In this case the Mexican woodland (the encinal and pine-oak woodland) species may have been involved. Vertical displacement of 300-400 m. would be sufficient to interconnect the woodland biotas of Southern Arizona with those of northern Mexico (119). The Mexican woodland species could have arrived during the glacio-pluvial maximum, but such a hypothesis presents the biotic difficulty of both a northern and a southern montane biota reaching their respective destributional limits at the same time. It is convenient to postulate northern invasion during a pluvial period of heavy winter precipitation, the typical environment of Cordilleran boreal biotas. The Madrean element of exclusively monsoon climatic preference is more likely to have moved north from Mexico during a warm wet interval.

Any paleoclimatic model of the Southwest must account for the dual character of the biota in the desert mountains. I regard the isolation of the populations of *Sceloporus jarrovii, S. scalaris, Crotalus lepidus, Quercus oblongifolia, Pinus engelmannii,* and many other Mexican montane elements as dating from the mid-postglacial, 4,000 to 8,000 B. P., when the distinctive summer wet climate of the Mexican Plateau was especially well developed.

A Wet Altithermal? To support his claim that the period of mid-postglacial arroyo cutting was arid, Antevs (6) has called attention to Hansen's mid-postglacial pollen record of high NAP in south-central Oregon. The pollen record of humid temperate regions elsewhere also features unmistakable evidence of hypsithermal (altithermal) warming (50).

Such is not the case in lower latitudes. In Hawaiian bogs Selling (161) found that rainforest elements (*Metrosideros, Myrsine, Cheirodendron, Cibotium,* and *Coprosma*) reached their maximum point of expansion at the expense of dry forest during the middle of postglacial time. "Although the European and North American mediocratic elements conclusively prove a higher temperature for Period II [the Hypsithermal] as compared with the others, and although the South American diagrams seem to show the same, only more abundant precipitation is indicated in the South Island of New Zealand [and Hawaii]."

In the arid Near East Murray (140) and Butzer (31) report that despite indications of higher local temperatures, the Neolithic moist interval featured a rainfall somewhat greater than that of the present. The Neolithic subpluvial interval is dated by Butzer at about 7,000 to 4,400 B. P., equivalent to the altithermal of Antevs.

There is meteorological justification for assuming that warm temperatures in high latitudes during the altithermal would enhance aridity in *winter* rain regions of California and the Great Basin. To extend the same paleoclimatic model to *summer* rain regions of the Southwest and the adjacent Mexican Plateau is to ignore the monsoon effect.

In the 20th century the summer rainfall of Mexico and southern Arizona was at a maximum from the turn of the century to the mid-1930's (176, 127). The gradual shifting northwards of the subtropical cells and thereby also of the middle latitude storm tracks during the first three decades of this century has long been known from studies of the continuous warming up in winter of the northern latitudes. "There is hardly any doubt that the increase of summer precipitation in Mexico during the same period is another proof for the shifting northwards of the subtropical high pressure areas and the middle latitude westerlies" (176).

Recognizing the significance of the summer monsoon in Southwestern paleoclimatology, Leopold and Miller (105) discuss its importance in their study of postglacial alluviation in Wyoming: "Summer in the Altithermal time was marked by low pressure over the continent, with a strong anti-cyclonic flow of air aloft around the upper air extension of the Bermuda high-pressure cell. As at present in the Southwest, considerable moisture derived from the Gulf of Mexico passed over the desert aloft, but the dearth of strong wave troughs aloft precluded the precipitation of that moisture except as infrequent but intense thunderstorms. Under such a climate, flash floods punctuating the hot, dry summers of the plains provided a period characterized by erosion of the valleys in the plains, but in the form of gullying rather than extensive sheet erosion."

In Arizona and New Mexico gully erosion in the altithermal represented the effect of summer convectional storms. Whether there was a significant increase in mean annual precipitation in the altithermal is less certain, but the increase in pine pollen in sediments associated with altithermal cutting suggests that the climate was more mesic than it is today.

What did the desert grassland look like in altithermal time? Increased summer rain would produce a richer cover of perennial grama grasses and bring the blue oaks, Emory oaks, and Chihuahua pines to lower elevations. The first thunderstorms of July struck a saffron yellow grassland, the creosote bushes bronze-colored, the ocotillo and coral bean leafless. Of the annuals, only hardy species occupying disturbed sites, such as prickly poppy and jimson weed, could be found in flower.

By the middle of August, after a series of cloudbursts, the landscape sparkled emerald, the color of a spring pasture on a Pennsylvania farm. Even then, despite exuberant plant growth, thunderstorms of high intensity, 8 cm. per hour, would yield 30 to 40 percent runoff to the channels. Intense channel erosion removed hundreds of tons of alluvium from the headwater flood plains to be spread across downstream flood plains. By early October the monsoon season had ended, the gramas, sacaton and needlegrass turned yellow again. Nine months of dry season lay ahead. The altithermal landscape of southern Arizona probably resembled the belt of grassland and encinal found today in Mexico east of the Sierra Madre. During the altithermal species of animals and plants in the Mexican encinal and grassland penetrated central Arizona and central New Mexico. At this time the early Cochise gatherers began cultivating corn under the favorable wet summer climate.

Summary. The classic Bryan-Antevs climatic model, deeply enmeshed in the literature of Southwestern prehistory and Pleistocene geology, relates prehistoric erosion to drought. I believe the same field evidence, with the addition of both the fossil pollen and the biogeographic record, supports a model in which postglacial erosion is attributed to periods of intense summer rainfall. Alluviation of the headwater flood plains occurred during periods of relatively light summer rainfall.

I find no reliable pollen evidence that postglacial droughts, if they occurred, were sufficient to shift biotic zones above their present level.

X. SOME CONCLUSIONS AND AFTERTHOUGHTS

Pollen analysis as a guide to climatic, cultural, and physiographic changes arose fifty years ago in Europe. Under the influence of von Post a systematic study of bog pollen profiles began; continuing studies throughout Europe and Russia have resulted in thousands of pollen diagrams from many hundreds of lakes, bogs, and archaeological sites. Pollen stratigraphy has become a cornerstone of Pleistocene stratigraphy in the Old World. The much richer flora, the more complex vegetation, and the lack of paleoecologists trained in pollen analysis has delayed the development of Pleistocene pollen stratigraphy in the New World. However, the time is long past for a more serious and concerted effort to apply pollen analysis in the development of paleoecology in the western hemisphere. Anthropologists, geologists, biogeographers and other interested in unwritten history will undoubtedly find the pollen record of use in reexamining their own findings and, as this study has proved to me, each of them has a crucial role to play in aiding the paleoecologist in his own explorations.

A singular advantage enjoyed by the pollen stratigrapher in the Southwest and in certain other parts of North America is the variety of undisturbed "climatic climax" plant communities. These serve as natural laboratories in which the scientist can examine processes which may illuminate the past. The Sonoran and Chihuahuan Deserts and the desert grassland still provide natural bench marks at which one may study the pollen rain of natural plant communities. By way of contrast the conversion of natural forest and savanna habitats in Europe began 4,000 years ago. Immersed in a mosaic of fields, pastures, managed woodlots, and villages, the European palynologist enjoys little opportunity to make critical comparisons between the fossil and modern pollen rain of natural areas.

But in the Southwest the time is short; man destroys as he builds, and under steady grazing by cattle, overly zealous fire protection, the inroads of agriculture, of mining, of military installations, and, most recently, the growth of subdivisions into the desert far from established cities, the natural landscape is being transformed. Behind the bulldozer blade Russian-thistle, London-rocket, puncture-vine and other introduced weeds spring up on the freshly broken ground, habitats denied to them when the soil is unbroken. In the absence of browsing animals native shrubs infest the closely grazed ranges. In the mountain forests under the fire lookout towers dense thickets of slow-growing saplings and shrubs shade out the fire-resistent bunch grass sod and choke the stately scattered pines of the 19th century. In the ore districts sterile tailings obliterate all vegetation. Along those flood plains not already converted to cotton fields a falling water table degrades natural cienegas to adobe flats and even mesquite may no longer be able to reach the water table with its long roots.

Despite these alarming changes, many natural areas will surely survive the present onrush of civilization. However, the pattern of change is not always predictable and the ecologist may be certain that a number of plant communities, whose value as natural laboratories he and other scientists are just beginning to appreciate, will be obliterated. At present it appears that natural flood plains and parts of the lower bajadas are the habitats most seriously endangered, with many areas already lost to agriculture. Even the undisturbed, uncultivated flood plain communities may be lost. The steady pumping of fossil water in adjacent farmland lowers the water table in valley bottoms beyond the point to which flood plain vegetation can reach it. Dead mesquite at Casa Grande Monument bear witness to the fact that pumping along the lower Santa Cruz and Gila Rivers has depleted the ground water reservoir formerly tapped by riparian trees and shrubs.

The ground water reservoir of the valley fill is a resource at least partly related to water storage during the Pleistocene. The economic necessity of mining ground water to produce cotton, a crop already in surplus, protected by tariff and supported by subsidy, poses a paradox most scientists would rather not confront. The legacy of modern flood plain agriculture will include a depleted water table, a harvest of useful but unessential food and fiber, and eventual economic crisis among the farming communities of the flood plains, and, as a byproduct, the destruction of the natural flood plain plant communities.

Until recently the upper bajadas appeared relatively immune from disturbance. Too rocky to be cultivated, too heterogeneous in rock types to be mined, lacking an adequate water supply for heavy cattle grazing, the upper bajada escaped intensive use. However, it affords a panorama of the desert mountains, a feature much prized by home owners and subdividers who are now busy bulldozing streets in quiet valleys where until recently only ranchers lived in ranch houses.

Near the top of Southwestern mountains small but biogeographically important and very vulnerable habitats exist — the scattered tiny populations of trees and other mesophytes on northeast exposures. In both the Catalina and Pinaleño Mountains the native stands of cork bark fir *(Abies lasiocarpa)* suffered needless damage when highway construction was routed through rather than around them. The natural pollen rain of these montane forests is of value in interpreting the natural pollen rain and the vegetation of the pluvial period.

Results of sampling the modern pollen rain forced me to shed some preconceived ideas. The pollen rain of the desert grassland is not a uniform mixture of all anemophilous species of plants. Local differences in community composition are immediately evident in both the fossil samples and in modern soil surface samples. While the cheno-ams (Goosefoot family plus *Amaranthus*) dominate the pollen record of most flood plains and cattle tanks on the lower bajada, the upper bajada contains a high frequency of composite pollen. Pollen of grass may be relatively abundant on either the flood plains (which support dense stands of the tall bunch grass *Sporobolus*) or on upper bajadas dominated by *Bouteloua* and *Aristida*. The mystery of what type of environment might have characterized the flood plains when they accumulated large quantities of low-spine composite pollen, as in pollen zones II and IV, was resolved by the discovery of a similar pollen rain in surface samples of a cienega beneath or near colonies of giant ragweed *(Ambrosia trifida)*. These matters might not have been easy to resolve if abundant "natural" plant communities were unavailable or if the local pollen rain was hopelessly mixed with pollen from disturbed habitats.

The objection that rebedding and redeposition will invalidate any ecological interpretation of flood plain pollen profiles can be discarded by examining the record. Pollen analysis reveals an orderly pattern in the counts, as one would expect from lake sediments. There is relatively little variation between adjacent levels and there are well-marked pollen zones. If rebedding or redeposition is deceiving us, it is doing so in an orderly fashion, rather than erratically or at ran-

dom. Long-distance transport of tree pollen, especially of pine and oak, is evident; these pollen types occur in livestock tanks and on the soil surface of the desert. The presence of a few percent of pine and oak pollen in alluvial deposits 10 to 50 kilometers from the nearest trees is largely the result of wind rather than water transport. But most of the pollen content of flood plain alluvium apparently comes largely from plants growing on the flood plain itself, rather than from the enclosing bajadas or uplands.

What ecological conditions are represented in the flood plain pollen record? I have pointed out that none of the arroyos I have studied in southern Arizona contain a pluvial-age pollen record, representing the period when pine forest or pine savanna invaded the desert grassland. Presumably none of the flood plain artifacts, including those of the Cochise Culture, should be considered contemporaneous with pluvial conditions; the entire record of the Cochise Culture can be related to minor variations in the present vegetation pattern. On cultural evidence, supported by radiocarbon dating, the Lehner and Naco sites appear to be somewhat older than the early Cochise Culture. Pollen study of the Lehner site is incomplete but there is no evidence to suggest that the 11,400 B. P. mammoth hunt took place under conditions vastly different from those typical of postpluvial times.

With rare exceptions the flood plain environment at present is dominated by cheno-am pollen. The cheno-ams reflect a low water table in the flood plain, a high soil salt content, and a climate characterized by heavy summer rains. Presumably this was the case 5,000 years ago when cheno-ams were dominant and when slightly higher frequencies of pollen of pine suggest a slightly wetter climate. Pollen zones II and IV dominated by composite pollen represent stability or alluviation in the flood plains with a high water table, relatively low soil salt accumulation, and a climate of fewer or lighter summer storms. Perhaps winter rains were heavier. If episodes of channel cutting and filling are essentially synchronous throughout the Southwest, as Bryan and others have maintained, it is likely that within the same vegetation zone the same plant communities will be present and the same type of local pollen rain will be sedimented in the separate flood plains of adjacent valleys. I hope it is clear that the assumption of synchronous cutting or filling throughout the years is crucial to what I have attempted — the building of a flood plain pollen chronology from different portions of the desert grassland. Should future studies indicate that there is an equal chance for either cheno-ams or composite pollen to be dominant in strata of the same age, with cutting under way in one valley and filling in the other, the matter of correlation will be much more difficult than I have imagined.

Prehistorians may be disappointed that the flood plain pollen record of *Zea* and other cultivated plants is relatively poor. It was possible to verify the presence of *Zea* through a series of levels only at the Point of Pines Cienega Creek site. Perhaps even under optimal conditions only relatively small portions of the flood plains were devoted to cultivation. By careful horizontal sampling it might be possible to map prehistoric corn fields. The fossil pollen record of *Zea* should help refine dating of the introduction of *Zea* into the Southwest. A much richer pollen record of economic plants can be found in rock shelters and in Pueblo ruins than in the alluvium of flood plains.

The archaeologist who would strengthen the stratigraphic control of his excavations can do so through pollen analysis. Flood plain sites contain a wealth of non-cultural information and merit close attention of both the geologist and ecologist. Flood plain sites are best treated as a paleo-environmental unit, as Antevs and Sayles attempted in their study of the Cochise Culture. Several pollen profiles at a single site would yield a far more precise control of the sedimentary sequence than I have been able to achieve my choosing widely scattered sites. And I have learned that to establish a firm correlation with archaeological finds, it is absolutely essential to collect pollen samples when the site is being excavated. To erect a ladder against an arroyo bank and to hack out a set of pollen samples from the hard walls is simple enough. But trying to relate the resulting pollen profile to the artifacts or bones previously collected at an excavation in the vicinity is never satisfactory, especially if the pollen sequence does not fully verify expectations.

The pollen record has led me to re-examine venerated paleo-climatic theory prevalent in the literature of Southwestern prehistory and climatology. The altithermal of 4,000 to 8,000 years ago is almost invariably spoken of as a time of drought. If biologically important droughts occurred in the Southwest in postpluvial time, I have failed to recognize them in the fossil record. Pollen evidence suggests that the altithermal was not hot and dry but rather relatively wet, at least in summer. Mexican species of plants and animals probably moved northward into southern Arizona at this time, to be isolated by subsequent climatic change and left as relicts in the desert mountains.

Whatever the nature of the altithermal, wet or dry, drought cannot be advanced as an explanation for the extinction of large mammals 8,000 to 10,000 years ago. The circular argument that drought caused extinction and that the presence of extinct animal bones is proof of wet climates can be discarded on empirical grounds. Extinct animal bones are to be found in sediments representing either pluvial (wet) or postpluvial (dry) environments.

Some anthropologists and paleontologists may dispute my insistence that large mammals disappeared not because they lost their food supply but because they became one. At first the suggestion that the earliest invaders of the New World were culturally so advanced and technically so skillful that they managed to destory more native species of animals than have fallen victim to the onslaught of western civilization seems preposterous. It violates our notion of cultural progress and our tendency to view Paleo-Indians as a part of the balance of nature. Perhaps we have understimated the population size and technical ability of the early hunters. Admittedly, our knowledge of Early Man is not terribly revealing despite the splendidly worked Clovis, Folsom, and other lanceolate projectile points. From the continent-wide distribution of the diagnostic projectile points Mason (126) has inferred a cultural homogeneity and a common base of subsistence based on fulltime hunting of large mammals. In the absence of any convincing alternative explanation the indictment of Early Man is unavoidable.

Following extinction of the large mammals the early hunters probably suffered economic depression and a population crash. Under a climate similar to the present and with the existing biotic zones in place, the early hunters were obliged to begin their 7,000-year experiment with native plants, leading in the altithermal to increasingly skillful techniques of harvesting and gathering, to the domestication of certain weedy camp-followers, and, within the last 1,000 years, to the widespread adoption of flood plain agriculture. Many clues along the trail remain to be detected by pollen analysis and other paleoecological methods.

Alluvium. Material deposited permanently or in transit by running water, including gravel, sand, silt, and clay; unconsolidated.

Altithermal (= *Long Drought*). An assumed warmer, drier climate than the present, during mid-postglacial time, 7,500 to 4,000 B. P. "Evidences are marked maxima of grass-chenopod-composite pollen in bogs in southcentral Oregon, practical disappearance of glaciers in the western mountains, drying up of lakes, arroyo — and wind — erosion, and calichification. Deserts attained considerably greater extent, occupying much of the now semi-arid regions" (6). Results of pollen analysis conflict with the notion of altithermal drought in the arid Southwest.

Anemophilous. Wind-pollinated. Such plants shed large quantities of pollen and may be "over-represented" in sediments.

AP. Arboreal pollen; sum of the tree and shrub pollen count.

Arroyo. The channel of an ephemeral or intermittent stream, usually with vertical banks of unconsolidated material two feet or more high (21).

Axial Stream. "The main stream of an intermontane valley, which flows along the lowest part of the valley and parallel to its longer dimension" (21).

Bajada. A gentle alluvial slope extending from mountain backwall to the inner valley flood plain. It is a surface of transport which may be either eroding or aggrading; the former are pediments, the latter are fans.

Biochore. A primary subdivision of the terrestrial environment, essentially characterized by the response of vegetation structure. Dansereau (44) recognizes forest, savanna, grassland, and desert.

Blancan Fauna. An early Pleistocene vertebrate fauna of North America preceded by an important wave of extinction (rhinoceroses, neohipparions, etc.), and characterized by the following genera: *Plesippus, Borophagus, Procastoroides,* and possibly *Mimomys.* Post Blancan faunas include *Equus, Bison,* and *Mammuthus* (128).

B. P. Before present; age in years before 1950.

Bryan-Antevs Model. A climatic interpretation of postglacial events in the Southwest which associates prehistoric arroyo cutting with drought and alluviation with more humid conditions.

Caliche. Gravel, sand, or clay cemented by calcium carbonate which forms secondarily in soils or alluvium of dry regions.

Cazador Stage. See Cochise Culture.

Channel Fill. Sediment deposited, often rapidly, in the bed of an arroyo.

Chaparral. A rather dense shrub vegetation, mostly evergreen, the tallest shrubs two to three meters high; in Arizona *Quercus turbinella* is a typical dominant.

Charco. "A hole in clay, or stratum of rock, where water collects, and from which it cannot run" (154).

Cheno-ams. Pollen of the goosefoot family (Chenopodiaceae) and the related genus *Amaranthus.* These were not distinguished in routine analysis.

Cheno-am/Composite Shift. A major change in pollen composition from dominance of composites to that of chenopods plus *Amaranthus* or vice versa.

Chihuahuan Desert. The driest portion of the Mexican Plateau, east of the Sierra Madre Occidental and including parts of Durango, San Luis Potosí, Coahuila, Chihuahua, trans-Pecos Texas, and southern New Mexico (Fig. 1); it is found largely above 1,000 m. Some ecologists would include southeastern Arizona. Characteristic shrubs are creosote, tarbush, mesquite, and ocotillo. Columnar cacti and many other arborescent species of the Sonoran Desert are absent.

Chiricahua Stage. See Cochise Culture.

Cienega. A natural fresh-water marsh with a high water table underlain by heavy aluvium and found occasionally along undissected flood plains. During arroyo cutting of the 19th century most Southwestern cienegas were drained.

Cochise Culture. The grinding tools and various stone implements of preceramic people in the Southwest. Four stages are recognized (152) in the following sequence from oldest to youngest: (1) Sulphur Spring Stage, an assemblage of grinding stones with flat metates, unshaped pebbles serving as manos, and few percussion-flaked tools; (2) Cazador Stage, moderately large basin metates, small, shaped manos, and the first appearance of flaked projectile points; (3) Chiricahua Stage, basin metates, shaped manos, percussion-flaked tools and pressure-flaked points, distinguished from the preceding by a greater frequency of flaked implements; (4) San Pedro Stage, deep basin metates, large manos, pressure-flaked projectile points, percussion-and pressure-flaked scrapers, axes and knives; pithouses and storage pits are diagnostic. Around 2,000 B.P. pottery appears.

Composites. The plant family Compositae; pollen in fossil sediment may be assigned to the following categories: *Artemisia,* Liguliflorae, high-spined and low-spined types. The low-spined Compositae include the genera *Ambrosia, Xanthium, Hymenoclea, Iva, Franseria,* and other wind-pollinated members of the tribe Heliantheae. The remainder of the composite genera are high-spined and not readily distinguished.

Desert Grassland = Mesquite Grassland. Short grass plains extending from Durango and Coahuila in the Mexican Plateau into southern Arizona, southern New Mexico, and southwestern Texas. Dominant genera of grasses include *Hilaria, Aristida, Bouteloua,* and *Sporobolus.* There are many annual forbs. In southern Arizona the vertical limits are roughly 1,000 to 1,500 m.

Dirt Tank. An artificial earth dam trapping runoff and used for watering livestock.

Dominant. A pollen component exceeding 10 percent of the total count.

Early Man. The New World big game hunters, best known through their finely flaked, lanceolate projectile points, e.g., Folsom, Clovis, Sandia.

Encinal. Woodland of low, open, round-crowned, scattered oaks, junipers, and pinyon, with grasses beneath. The trees are 3 to 10 m. in height. Not to be confused with chaparral, which is more dense, generally much lower in height, and lacking in grasses.

Hypsithermal. The warmer part of the postglacial period; according to Deevey and Flint (62) it lasted from 9,500 to 2,500 B. P.

Inner-valley Alluvium. Flood plain deposits of axial streams at the lowest point of a valley, inset in older and usually coarser valley fill. From the radiocarbon dates it appears the inner valley alluvium is typically of postglacial age (10,000 B. P., or younger).

Late-glacial. The period between the end of the Wisconsin glacial maximum of 17,000 years ago and the start of climatic conditions subequal to the present, 10,000 years ago. In Europe it is a time when tundra and birch woodland invaded freshly deglaciated terrain before the arrival of forest. In the Southwest late-glacial age pollen deposits are reported in the San Juan Mountains and in the Llano Estacado.

Long Drought. (See Altithermal.) The term applied by Antevs to the interval of intense arroyo cutting, calichification, and, in certain areas, of dune formation during the mid-postglacial.

Merriam Effect. Mountain height and mass influence vertical distribution of vegetation. Mesic plants will descend to a lower level on the more massive, higher range (134, 110). Van Steenis (170) describes this phenomenon in Java and refers to it by its European term "Massenerhebung." I prefer to follow Lowe in using the term Merriam effect.

Mesophytes. Plants requiring constant water availability, e.g. ash, fir, and grape.

Mesquite Mounds. Dune-like hills of sand and silt enveloping roots and stems of *Prosopis velutina torreyana.*

Metal Rim. A common form of stock tank above the reach of cattle; it serves as an excellent trap of atmospheric pollen.

Mexican Plateau. The arid upland of basins and ranges rising in elevation from 1,200 m. in southern Arizona and New Mexico (lat. 33° north) to 2,100 m. near Mexico City (lat. 19°

north). It is bounded by the Sierra Madre Occidental, Sierra Madre Oriental, Rio Grande Valley, and the Colorado Plateau. Biotically the region is recognized as a center of speciation for many of the major North American plant and animal taxa.

Monsoon. Anticyclonic summer precipitation; in the Southwest the heaviest rainfall occurs in July and August.

NAP. Non-arboreal pollen; sum of the forb, grass, and sedge pollen count.

Over-representation. The tendency for plants growing near the site of deposition to contribute a disproportionate amount of pollen, e.g. sedges and grasses may be over-represented in sediment of mountain parks.

Paleo-Indians. See Early Man.

Palynology. The study of pollen and spores.

Parkland. A savanna of especially tall trees (15-30 m.)

Pediment. A gently sloping erosional surface bevelling either bedrock or valley fill.

Phreatophyte. An indicator of a shallow water table (130).

Pine-oak Woodland. A position in the Southwestern vegetation gradient lying immediately below ponderosa pine forest and featuring mixtures of Chihuahua pine, Apache pine, Arizona oak, Emory oak, and silverleak oak (115).

Pink Caliche. The top of the valley fill occasionally exposed beneath the present inner-valley flood plain in parts of the Southwest; a term used by Sayles and Antevs (151).

Pithouse. A prehistoric dwelling place dug a meter or more into the ground, especially along flood plains. Pithouses were part of the San Pedro and subsequent cultures.

Playa. The dry bed of an ancient lake, generally of pluvial age.

Pleistocene Megafauna. Large vertebrates of the glacial period which disappeared relatively recently, e.g. *Mammuthus, Equus, Nothrotherium,* and *Camelops.*

Pluvial. An interval when the now dry lakes of the Southwest held permanent water and when the climate was considerably more moist. Correlation of Great Basin pluvials and Wisconsin glacial events has been demonstrated by radiocarbon dating (62, 63). Bryan (27) argues in behalf of avoiding the term "pluvial" with its implication of wetter conditions and more violent rains. "Glacial" is a recommended synonym.

Pollen Component. The pollen of a particular genus or family or other taxonomic unit which can be readily identified and plotted in a pollen profile, e.g. pines, grasses, composites, etc.

Postglacial = Postpluvial. The last 10,000 years. Flint (62) states that ". . . the term postglacial is hardly applicable to any feature in the nonglaciated district." In the arid southwest a "pluvial climate" prevailed at various times in the past, ending around 11,000 years ago. Postpluvial indicates the climatic condition not greatly different from the present climate which has endured in the last 10,000 years. "Recent" is the official U. S. Geological Survey term for part of the last 10,000 years.

It would be convenient to have a time-stratigraphic term to apply to the inner valley alluvium exposed in western arroyos. I cannot use Recent as it excludes deposits containing extinct animals (137). "Postpluvial alluvium" may be suitable for the moment. From both the pollen record and from radiocarbon dates it appears that the deposits under study postdate the existence of permanent water in the playa lakes of western North America.

Preceramic. Archaeological remains predating the adoption of pottery traits; in the Southwest pottery became widespread about 2,000 years ago.

Riparian. Of the shore; e.g. cottonwood and ash woods along river flood plains in the desert.

San Pedro Stage. See Cochise Culture.

Savanna. An open, scattered growth of trees, for example, blue oaks in grama grassland; pinyon-juniper woodland; the subarctic taiga with widely spaced black spruce. As employed by Dansereau the term is of structural significance exclusively. Grassland lacks trees (except along rivers); forest presents a closed canopy; the savanna is an intermediate condition.

Sonoran Desert. The extremely arid region of western Sonora, most of Baja California, extreme southwestern California, and southern and western Arizona below 900-1,200 m. (see Fig. 3). It is characterized by a rich arborescent flora including large columnar (e.g. saguaro, organ pipe) cacti on upper slopes and the ubiquitous creosote bush on the intermontane valleys and plains.

Southwest (United States). Arizona, New Mexico, and adjacent arid parts of Texas, Utah, Colorado, southern Nevada and eastern California, a region of distinctive biotic, climatic, and cultural conditions. Anthropologists and ecologists long accustomed to this definition may be surprised to learn that a recent book on woody plants of the Southwest includes eastern Texas, Louisiana, Oklahoma and Arkansas and excludes Arizona, southern Utah and Nevada.

Stock Tank. An artificial storage basin supplying drinking water to livestock. Two common types are eight-meter diameter "metal-rims," supplied by wells, and "dirt tanks," supplied by flood water runoff.

Sulphur Spring Stage. See Cochise Culture.

Valley Fill. Late Cenozoic alluvial waste or lacustrine sediments filling to a considerable depth basins of the Southwest. Bedrock may be over 1,000 m. below the surface of the fill.

Well-cutting. A mud chip caught in the bit and brought to the surface in a well drilling.

Wisconsin Age. The last glacial period at the end of the Pleistocene, dated at roughly 35,000 to 10,000 B.P.

Woodland. A savanna with trees of medium height, 3-15 m.

Xerophytes. Perennial plants able to withstand prolonged drought, e.g., ocotillo, creosote bush, yucca.

Zoophilous. Animal-pollinated, usually by insects. Pollen of such plants is poorly represented or absent in sediments.

SCIENTIFIC AND COMMON NAMES OF PLANTS AND ANIMALS MENTIONED IN TEXT

(a = animal- or self-pollinated; w = wind-pollinated)

	Name	Family	Other name
w	*Abies*	Pinaceae	fir
w	*Abies concolor*	Pinaceae	white fir
w	*Abies lasiocarpa*	Pinaceae	alpine fir
a	*Abronia*	Amaranthaceae	sand verbena
a	*Abutilon*	Malvaceae	Indian mallow
a	*Acacia*	Leguminosae	acacia
a	*Acacia constricta*	Leguminosae	white-thorn
a	*Acacia constricta vernicosa*	Leguminosae	
a	*Acalypha neomexicana*	Euphorbiaceae	
a	*Acer glabrum*	Aceraceae	Rocky Mountain maple
a	*Acer grandidentatum*	Aceraceae	big-tooth maple
a	*Agave*	Amaryllidaceae	
a	*Agave palmeri*	Amaryllidaceae	century plant
w	Alder	Betulaceae	*Alnus*
a	*Allionia*	Nyctaginaceae	
a	*Allionia incarnata*	Nyctaginaceae	trailing four-o'clock
w	*Alnus*	Betulaceae	alder
a	*Alternanthera repens*	Amaranthaceae	
w	*Amaranthus*	Amaranthaceae	amaranth
w	*Amaranthus graecizans*	Amaranthaceae	
w	*Amaranthus palmeri*	Amaranthaceae	careless-weed, red-root
w	*Ambrosia*	Compositae	ragweed
w	*Ambrosia psilostachya*	Compositae	perennial ragweed
w	*Ambrosia trifida*	Compositae	giant ragweed
a	*Apodanthera undulata*	Cucurbitaceae	melon-loco
a	*Arbutus arizonica*	Ericaceae	Arizona madroño
a	*Arctostaphylos*	Ericaceae	manzanita
a	*Arctostaphylos pringlei*	Ericaceae	
a	*Arctostaphylos pungens*	Ericaceae	point-leaf manzanita
w	*Aristida*	Gramineae	three-awn
w	*Aristida adscensionis*	Gramineae	six-weeks three-awn
w	*Aristida divaricata*	Gramineae	poverty three-awn
w	*Aristida glabrata*	Gramineae	
w	*Aristida longiseta*	Gramineae	red three-awn
w	*Aristida ternipes*	Gramineae	spider grass
w	*Artemisia*	Compositae	sagebrush, wormwood
w	*Artemisia dracunculoides*	Compositae	false tarragon
w	*Artemisia filifolia*	Compositae	sand sagebrush
w	Ash	Oleaceae	*Fraxinus*
a	*Aster tanacetifolius*	Compositae	
w	*Atriplex*	Chenopodiaceae	salt-bush, orache
a	*Baccharis*	Compositae	
a	*Baccharis glutinosa*	Compositae	seep-willow, batamote
a	*Bahia*	Compositae	
a	*Bahia absinthifolia*	Compositae	
a	*Baileya multiradiata*	Compositae	desert marigold
w	Basswood	Tiliaceae	*Tilia*
a	*Berlandiera lyrata*	Compositae	
w	*Betula*	Betulaceae	Birch
w	Birch	Betulaceae	*Betula*
a	*Boerhaavia coulteri*	Nyctaginaceae	spiderling
a	*Boerhaavia erecta*	Nyctaginaceae	spiderling
a	*Boerhaavia intermedia*	Nyctaginaceae	spiderling
a	*Boerhaavia wrightii*	Nyctaginaceae	spiderling
w	*Bouteloua*	Gramineae	grama
w	*Bouteloua aristidoides*	Gramineae	needle grama
w	*Bouteloua barbata*	Gramineae	six-weeks grama
w	*Bouteloua chondrosioides*	Gramineae	spruce-top grama
w	*Bouteloua curtipendula*	Gramineae	side-oats grama
w	*Bouteloua eriopoda*	Gramineae	black grama
w	*Bouteloua filiformis*	Gramineae	slender grama
w	*Bouteloua gracilis*	Gramineae	blue grama
w	*Bouteloua hirsuta*	Gramineae	hairy grama
w	*Bouteloua rothrockii*	Gramineae	rothrock grama
a	Burroweed	Compositae	*Haplopappus tenuisectus*
a	Cactus	Cactaceae	
a	*Calliandra*	Leguminosae	False mesquite
a	*Calliandra humilis*	Leguminosae	
a	*Carnegiea gigantea*	Cactaceae	saguaro
w	*Carpinus*	Betulaceae	hornbeam
w	*Carya*	Juglandaceae	hickory
a	*Cassia bauhinioides*	Leguminosae	
w	Cat-tail	Typhaceae	*Typha*
w	*Celtis*	Ulmaceae	hackberry
w	*Celtis pallida*	Ulmaceae	desert hackberry
	Cheirodendron	Araliaceae	
w	Chenopodiaceae		goose-foot family
w	*Chloris*	Gramineae	
w	*Chloris virgata*	Gramineae	feather-finger grass
a	Cholla, cane	Cactaceae	*Opuntia fulgida*
w	*Cibotium*	Cyatheaceae	
a	*Citharexylum brachyanthum*	Verbenaceae	
a	*Coldenia greggii*	Boraginaceae	
a	*Commelina erecta*	Commelinaceae	day flower
a	*Condalia lycioides*	Rhamnaceae	gray-thorn
a	*Condalia spathulata*	Rhamnaceae	squaw-bush
w	*Coprosma*	Rubiaceae	
a	Coral bean	Leguminosae	*Erythrina flabelliformis*
w	Corn	Gramineae	*Zea*
a	Creosote bush	Zygophyllaceae	*Larrea tridentata*
	Crotalus lepidus	Crotalidae	rock rattlesnake
a	*Croton*	Euphorbiaceae	croton
a	*Croton corymbulosus*	Euphorbiaceae	
a	*Cucurbita foetidissima*	Cucurbitaceae	buffalo gourd
a	*Cupressus*	Cupressaceae	cypress
a	*Dasylirion wheeleri*	Liliaceae	sotol
w	Douglas fir	Pinaceae	*Pseudotsuga menziesii* (to replace *P. taxifolia*)
a	*Drymaria sperguloides*	Caryophyllaceae	sand spurry
w	Elm	Ulmaceae	*Ulmus*
a	*Encelia*	Compositae	brittle-bush
w	*Ephedra*	Ephedraceae	joint-fir, Mormon tea
w	*Ephedra antisyphilitica*	Ephedraceae	
w	*Ephedra aspera*	Ephedraceae	
w	*Ephedra californica*	Ephedraceae	California ephedra
w	*Ephedra clokeyi*	Ephedraceae	
w	*Ephedra coryi*	Ephedraceae	
w	*Ephedra funera*	Ephedraceae	
w	*Ephedra nevadensis*	Ephedraceae	Nevada ephedra
w	*Ephedra torreyana*	Ephedraceae	Mormon tea
w	*Ephedra trifurca*	Ephedraceae	Mexican tea
w	*Ephedra viridis*	Ephedraceae	green ephedra
w	*Eragrostis arida*	Gramineae	love grass
w	*Eragrostis megastycha*	Gramineae	stink grass
w	*Eriochloa lemmoni*	Gramineae	cup grass
a-w	*Eriogonum*	Polygonaceae	wild buckwheat
w	*Eurotia*	Chenopodiaceae	winter-fat
a	*Eupatorium greggii*	Compositae	
a	Euphorbiaceae		spurge family
a	*Euphorbia*	Euphorbiaceae	spurge
a	*Euphorbia albomarginata*	Euphorbiaceae	
a	*Euphorbia capitellata*	Euphorbiaceae	
a	*Euphorbia revoluta*	Euphorbiaceae	
a	*Euphorbia serpyllifolia*	Euphorbiaceae	
a	*Euphorbia serrula*	Euphorbiaceae	
a	*Euphorbia stictospora*	Euphorbiaceae	
a	*Evolvulus*	Convolvulaceae	
w	Fir	Pinaceae	*Abies*
a	*Flourensia*	Compositae	tar-bush, varnish-bush
a	*Flourensia cernua*	Compositae	tar-bush, varnish-bush
a	*Fouquieria splendens*	Fouquieriaceae	ocotillo
w	*Franseria*	Compositae	bur-sage

w	*Franseria acanthicarpa*	Compositae	bur-weed
w	*Fraxinus*	Oleaceae	ash
a	*Froelichia gracilis*	Amaranthaceae	snake-cotton
a	*Gaillardia pulchella*	Compositae	fire-wheel, Indian blanket
a	*Garrya wrightii*	Cornaceae	silk tassel
a	*Gaura parviflora*	Onagraceae	
a	*Gayoides*	Malvaceae	
	Geomys	Geomyidae	pocket gopher
a	*Gilia longiflora*	Polemoniaceae	
a	*Gossypium*	Malvaceae	cotton
w	Grama	Gramineae	*Bouteloua*
a	*Gutierrezia*	Compositae	snakeweed
a	*Haplopappus*	Compositae	
a	*Haplopappus laricifolius*	Compositae	turpentine-brush
a	*Haplopappus tenuisectus*	Compositae	burroweed
a	*Helenium hoopesii*	Compositae	orange sneeze-weed
w	Hemlock	Pinaceae	*Tsuga*
a	*Hibiscus*	Malvaceae	rose-mallow
a	*Hibiscus denudatus*	Malvaceae	rose-mallow
w	*Hilaria*	Gramineae	
w	*Hilaria belangeri*	Gramineae	curly-mesquite
w	*Hilaria mutica*	Gramineae	tobosa
a	*Hoffmanseggia densiflora*	Leguminosae	hog-potato
a	*Horsfordia*	Malvaceae	
w	*Hymenoclea*	Compositae	burro-brush
a	*Hymenothrix wrightii*	Compositae	
a	*Ipomoea hirsutula*	Convolvulaceae	morning glory
a	*Iris missouriensis*	Iridaceae	Rocky Mountain iris
a	*Jatropha macrorhiza*	Euphorbiaceae	
a	Jimson weed	Solanaceae	*Datura meteloides*
w	Johnson grass	Gramineae	*Sorghum halepense*
w	Juglans	Juglandaceae	walnut
w	Juniper	Cupressaceae	*Juniperus*
w	*Juniperus*	Cupressaceae	juniper
w	*Juniperus deppeana*	Cupressaceae	alligator juniper
a	*Kallstroemia*	Zygophyllaceae	
a	*Kallstroemia californica*	Zygophyllaceae	
a	*Kallstroemia grandiflora*	Zygophyllaceae	Arizona poppy, Mexican poppy
a	*Koeberlinia spinosa*	Koeberliniaceae	Crucifixion-thorn
a	*Larrea*	Zygophyllaceae	creosote-bush
a	*Larrea tridentata*	Zygophyllaceae	creosote-bush
a	*Lepidium* sp.	Cruciferae	pepper-grass
a	*Lepidium thurberi*	Cruciferae	
a	*Lippia wrightii*	Verbenaceae	
a	London rocket	Cruciferae	*Sisymbrium irio*
a	*Lycium berlandieri*	Solanaceae	wolf-berry
a	Martyniaceae		unicorn-plant family
a	*Menodora*	Oleaceae	
a	*Menodora scabra*	Oleaceae	
a	*Mentzelia pumila*	Loasaceae	stick-leaf
a	Mesquite	Leguminosae	*Prosopis*
a	*Metrosideros*	Myrtaceae	
a	*Microrhamnus ericoides*	Rhamnaceae	
a	*Mirabilis*	Nyctaginaceae	four-o'clock
a	*Mollugo verticillata*	Aizoaceae	carpet-weed
w	*Muhlenbergia*	Gramineae	muhly
w	*Muhlenbergia porteri*	Gramineae	bush muhly
w	*Myrsine*	Myrsinaceae	
a	*Nolina microcarpa*	Liliaceae	sacahuista, bear-grass
a	Nyctaginaceae		four-o'clock family
w	Oak	Fagaceae	*Quercus*
w	Oak, coast-live	Fagaceae	*Quercus agrifolia*
w	Oak, emory	Fagaceae	*Quercus emoryi*
w	Oak, Mexican blue	Fagaceae	*Quercus oblongifolia*

a	Ocotillo	Fouquieriaceae	*Fouquieria splendens*
a	*Opuntia*	Cactaceae	prickly pear and cholla
a	*Opuntia imbricata*	Cactaceae	cholla
a	*Opuntia leptocaulis*	Cactaceae	Christmas cactus
w	*Ostrya*	Betulaceae	hop-hornbeam
a	*Oxybaphus*	Nyctaginaceae	
a	Palo verde	Leguminosae	*Cercidium*
w	*Panicum obtusum*	Gramineae	vine-mesquite
w	*Parthenium*	Compositae	mariola
a	*Parthenium incanum*	Compositae	mariola
a	*Pectis filipes*	Compositae	
a	*Pectis prostrata*	Compositae	
a	*Perezia wrightii*	Compositae	
a	*Physalis wrightii*	Solanaceae	ground-cherry
w	*Picea*	Pinaceae	spruce
w	*Picea engelmanni*	Pinaceae	Engelmann spruce
w	Pine	Pinaceae	*Pinus*
w	Pine, Chihuahua	Pinaceae	*Pinus leiophylla*
w	Pine, jack	Pinaceae	*Pinus banksiana*
w	Pine, western yellow	Pinaceae	*Pinus ponderosa*
w	*Pinus aristata*	Pinaceae	bristle-cone pine
w	*Pinus banksiana*	Pinaceae	jack pine
w	*Pinus cembroides*	Pinaceae	Mexican pinyon
w	*Pinus chihuahuana*	Pinaceae	Chihuahua pine
w	*Pinus durangensis*	Pinaceae	
w	*Pinus edulis*	Pinaceae	Colorado pinyon
w	*Pinus engelmannii*	Pinaceae	Engelmann pine
w	*Pinus latifolia*	Pinaceae	Apache pine
w	*Pinus leiophylla*	Pinaceae	Chihuahua pine
w	*Pinus monophylla*	Pinaceae	single-leaf pinyon
w	*Pinus ponderosa scopulorum*	Pinaceae	western yellow pine
w	*Pinus ponderosa arizonica*	Pinaceae	western yellow pine
w	*Pinus strobiformis* or *P. reflexa*	Pinaceae	southwestern white pine
w	Pinyon	Pinaceae	*Pinus*
a-w	*Plantago*	Plantaginaceae	plantain, Indian wheat
w	*Platanus*	Platanaceae	sycamore
a-w	*Polygonum*	Polygonaceae	knotweed, smartweed
a	*Polygonum coccineum*	Polygonaceae	
w	*Populus*	Salicaceae	poplar
w	*Populus tremuloides*	Salicaceae	quaking aspen
a	*Portulaca umbraticola*	Portulacaceae	purslane
a	*Proboscidea parviflora*	Martyniaceae	unicorn-plant
a	*Prosopis*	Leguminosae	mesquite
a	*Prosopis juliflora torreyana*	Leguminosae	mesquite
a	*Prosopis juliflora velutina*	Leguminosae	mesquite
w	*Pseudotsuga*	Pinaceae	Douglas fir
w	*Pseudotsuga menziesii* or *P. taxifolia*	Pinaceae	Douglas fir
a	Puncture-vine	Zygophyllaceae	*Tribulus terrestris*
w	*Quercus arizonica*	Fagaceae	Arizona white oak
w	*Quercus emoryi*	Fagaceae	Emory oak
w	*Quercus gambelii*	Fagaceae	Gambel oak
w	*Quercus hypoleucoides*	Fagaceae	silverleaf oak
w	*Quercus oblongifolia*	Fagaceae	Mexican blue oak
w	*Quercus reticulata* or *rugosa*	Fagaceae	net-leaf oak
w	*Quercus toumeyi*	Fagaceae	Toumey oak
w	*Quercus turbinella*	Fagaceae	shrub live oak
w	Ragweed, giant	Compositae	*Ambrosia trifida*
w	Ragweed, perennial	Compositae	*Ambrosia psilostachya*
a	*Rhus microphylla*	Anacardiaceae	Littleleaf sumac
a	*Robinia neo-mexicana*.	Leguminosae	New Mexican locust
w	Russian-thistle	Chenopodiaceae	*Salsola kali*
w	Sacaton	Gramineae	*Sporobolus wrightii*
w	Sagebrush	Compositae	*Artemisia*
w-a	*Salix*	Salicaceae	willow

74

w-a	*Salix scouleriana*	Salicaceae	scouler willow
a	*Sanvitalia aberti*	Compositae	
w	*Sarcobatus*	Chenopodiaceae	grease-wood
	Sceloporus jarrovii	Iguanidae	Yarrow's lizard
	Sceloporus scalaris	Iguanidae	bunchgrass lizard
w	*Scirpus*	Cyperaceae	tule, bulrush
w	*Scleropogon*	Gramineae	burro-grass
w	*Scleropogon brevifolius*	Gramineae	burro-grass
a	*Senecio* sp.	Compositae	groundsel
a	*Sericodes gregii*	Zygophyllaceae	
w	*Setaria macrostachya*	Gramineae	plains bristle grass
a	*Sida leprosa*	Malvaceae	
a	*Sida procumbens*	Malvaceae	
a	*Sidalcea*	Malvaceae	checker-mallow
a	*Simmondsia*	Buxaceae	deer-nut, jojoba, coffee-bush
a	*Solanum rostratum*	Solanaceae	buffalo-bur
	Sonorella	Xanthonychidae	Sonoran snail
w	*Sorghum halepense*	Gramineae	Johnson grass
w	*Sporobolus*	Gramineae	drop-seed
w	*Sporobolus airoides*	Gramineae	alkali-sacaton
w	*Sporobolus wrightii*	Gramineae	sacaton
w	*Spruce*	Pinaceae	*Picea*
w	*Suaeda*	Chenopodiaceae	seep-weed, quelite-salado
a	*Talinum aurantiacum*	Portulacaceae	
	Thomomys botteae	Geomyidae	pocket gopher
a	*Tidestromia*	Amaranthaceae	
a	*Tidestromia lanuginosa*	Amaranthaceae	
a	*Tilia*	Tiliaceae	linden
a	*Trianthema portulacastrum*	Aizoaceae	pigweed
w	*Trichachne californica*	Gramineae	cotton-top
w	*Trichostema arizonicum*	Labiatae	blue curl
w	*Tridens muticus*	Gramineae	
w	*Tridens pulchellus*	Gramineae	
a	*Tripterocalyx*	Nyctaginaceae	
w	*Tsuga*	Pinaceae	hemlock
w	*Tule*	Cyperaceae	*Scirpus*
w	*Typha*	Typhaceae	cat-tail
w	*Ulmus*	Ulmaceae	elm
a	*Vauquelinia californica*	Rosaceae	Arizona rosewood
a	*Verbesina encelioides*	Compositae	crown-beard
w	*Walnut*	Juglandaceae	*Juglans*
a	*White-thorn*	Leguminosae	*Acacia constricta*
a-w	*Willow*	Salicaceae	*Salix*
w	*Xanthium*	Compositae	cocklebur
w	*Xanthium saccharatum*	Compositae	cocklebur
a	*Yucca*	Liliaceae	soap-weed, spanish-bayonet, datil
a	*Yucca australis*	Liliaceae	
a	*Yucca elata*	Liliaceae	soap-tree yucca, palmilla
a	*Yucca schottii*	Liliaceae	mountain yucca
a	*Yucca torreyi*	Liliaceae	torrey yucca
w	*Zea*	Gramineae	corn, maize
a	*Zinnia*	Compositae	zinnia
a	*Zinnia grandiflora*	Compositae	
a	*Zinnia pumila*	Compositae	

LITERATURE CITED

1. Albritton, C. C., Jr. and K. Bryan. 1939. Quaternary stratigraphy in the Davis Mountains, Trans-Pecos, Texas. Bull. Geol. Soc. Amer. 50:1423-1474.
2. Andersen, S. T. 1954. A late-glacial pollen diagram from southern Michigan, U. S. A. Danm. Geol. Unders., II Raekke 80:140-155.
3. Antevs, E. 1952. Arroyo-cutting and filling. Jour. Geol. 60:375-385.
4. ———— 1954. Climate of New Mexico during the last Glacio-Pluvial. Jour. Geol. 62:182-191.
5. ———— 1955a. Geologic-climatic dating in the west. Amer. Antiquity 20:317-355.
6. ———— 1955b. Geologic-climatic method of dating. *Geochronology*, T. L. Smiley, ed. Univ. Ariz. Phys. Sci. Bull. No. 2:151-169.
7. ———— 1959. Geological age of the Lehner mammoth site. Amer. Antiquity 25:31-34.
8. Arellano, A. R. V. 1953. Barrilaco pedocal, a stratigraphic marker ca. 5,000 B. C. and its climatic significance. *Deserts actuels et anciens.* Congres Geologique International, Sec. VII, pp. 53-76.
9. Arms, B. C. 1960. A silica depressant method for concentrating fossil pollen and spores. Micropaleontology 6:327-328.
10. Aschmann, H. H. 1958. Great Basin climates in relation to human occupance. Univ. California Archaeological Survey Reports 42:23-40.
11. Benson, L. and R. A. Darrow. 1954. *Trees and Shrubs of the Southwestern Deserts.* Univ. Ariz. Press, Tucson. 437 pp.
12. Bent, A. M. 1960. Pollen analysis at Deadman Lake, Chuska Mountains, New Mexico. M.S. thesis. University of Minnesota, 22 pp.
13. Blumer, J. S. 1910. A comparison between two mountain sides. Plant World 13:134-140.
14. ———— 1911. Change of aspect with altitude. *Ibid.* 14:236-248.
15. Boughey, A. S. 1955. The vegetation of the mountains of Biafra. Pro. Linn. Soc. London 165:144-150.
16. Boyko, H. 1947. On the role of plants as quantitative climate indicators and the geo-ecological law of distribution. Jour. Ecology 35:138-157.
17. Brand, D. D. 1936. Notes to accompany a vegetation map of northwest Mexico. Univ. New Mex. Bull. No. 280, biol. series 4 (4), 27 pp.
18. Branscomb, B. L. 1958. Shrub invasion of a southern New Mexico desert grassland range. Jour. Range Management 11:129-132.
19. Bretz, J. H. and L. Horberg. 1949. Caliche in southeastern New Mexico. Jour. Geology 57:491-511.
20. Brown, C. N. 1956. The origin of caliche on the northeastern Llano Estacado, Texas. Jour. Geology 64:1-15.
21. Bryan, K. 1923. Erosion and sedimentation in Papago country, Arizona. U.S.G.S. Bull. 730.
22. ———— 1925. Date of channel trenching (arroyo cutting) in the arid southwest. Science 62:338-344.
23. ———— 1928. Change in plant associations by change in ground water level. Ecology 9:474-478.
24. ———— 1940. Erosion in the valleys of the Southwest. New Mex. Quarterly 10:227-232.
25. ———— 1941. Pre-Columbian agriculture in the Southwest as conditioned by periods of alluviation. Ann. Assoc. Amer. Geographers 31:219-242.
26. ———— 1948. Los suelos complejos y fósiles de la altiplanicie de México, en relación a los cambios climaticos. Bol. Soc. Geol. Mex. 13:1-20.
27. ———— 1950. Geological interpretation. *In* E. W. Haury, *Ventana Cave.* Univ. Ariz. Press, pp. 75-126.
28. ———— 1954. The geology of Chaco Canyon, New Mexico, in relation to the life and remains of the prehistoric peoples of Pueblo Bonito. Smithsonian Misc. Coll. 122 (7):1-65.
29. ———— and C. C. Albritton, Jr. 1943. Soil phenomena as evidence of climatic changes. Amer. Jour. Sci. 241:469-490.

30. Bryson, R. A. and W. P. Lowry. 1955. Synoptic climatology of the Arizona summer precipitation singularity. Bull. Amer. Meteorol. Soc. 36:329-339.

31. Butzer, K. W. 1958. Quaternary statigraphy and climate in the Near East. Bonner Geographische Abhandlungen 24:1-157.

32. Campbell, R. S. and I. F. Campbell. 1938. Vegetation on gypsum soils of the Jornada plain, New Mexico. Ecology 19:572-577.

33. Castetter, E. F. 1956. The vegetation of New Mexico. New Mexico Quarterly 26:257-288.

34. Clisby, K. H., F. Foreman and P. B. Sears. 1957. Pleistocene Climatic Changes in New Mexico, U. S. A. Veröff. Geobotanisches Institut Rübel (Zürich) 34:21-26.

35. ——— and P. B. Sears. 1956. San Augustin Plains — Pleistocene climatic changes. Science 124:537-539.

36. Coates, D. R. and R. L. Cushman. 1955. Geology and groundwater resources of the Douglas Basin, Arizona. Geol. Surv. Water-supply Paper 1354.

37. Colton, H. S. 1932. Sunset Crater, the effect of a volcanic eruption on an ancient people. Geog. Rev. 22:582-590.

38. Cooper, C. F. 1957. The variable plot method for estimating shrub density. Jour. Range Management 10:111-115.

39. ——— 1960. Changes in vegetation, structure, and growth of Southwestern pine forests since White settlement. Ecol. Monog. 30:129-164.

40. Culley, M. 1943. Grass grows in the summer or not at all. Hereford Jour., Sept. 1, 1943.

41. Damon, Paul E. 1962. Correlation and chronology of ore deposits and volcanic rocks. Annual Progress Report No. 4. Mimeo.

42. ——— and Austin Long. 1962. Arizona radiocarbon dates III. Radiocarbon 4:239-249.

43. ——— and J. Sigalove. MS. Arizona radiocarbon dates IV.

44. Dansereau, P. 1957. *Biogeography an Ecological Perspective.* Ronald Press Co., New York. 394 pp.

45. Darrow, R. A. 1944. Arizona range resources and their utilization. I. Cochise County. Ariz. Agr. Exp. Sta. Bull. 103:311-363.

46. Darwin, C. 1855. *Journal of Researches.* Harper & Bros., New York.

47. Daubenmire, R. F. 1938. Merriam's life zones of North America. Quat. Rev. Biol. 13:327-332.

48. ——— 1946. The life zone problem in the northern intermountain region. Northwest Science 20:28-38.

49. Deevey, E. S. 1957. Radiocarbon-dated pollen sequences in eastern North America. Veröff. Geobotanisches Institut. Rübel (Zürich) 34:30-37.

50. ——— and R. F. Flint. 1957. Postglacial hypsithermal interval. Science 125:182-184.

51. Deraniyagala, P. E. P. 1955. Some exinct elephants, their relatives and the two living species. Ceylon National Museums Publ. 161 pp.

52. Dimbleby, G. W. 1957. Pollen analysis of terrestrial soils. New Phytologist 65:12-28.

53. Dorroh, J. H. 1946. Certain hydrologic and climatic characteristics of the Southwest. Univ. New Mexico Publ. in Engineering No. 7:1-64.

54. Dutton, C. E. 1882. Tertiary history of the Grand Canyon district. U. S. Geol. Survey Mono. 2. 269 pp.

55. Eddy, F. 1958. Sequence of cultural and alluvial deposits in the Cienega Creek basin, southeastern Arizona. Univ. Ariz. M.S. thesis.

56. Ellison, L. 1951. Indicators of condition and trend on high range-watersheds of the intermontane region. U. S. Dept. Agriculture Handbook No. 19:1-66.

57. ——— 1954. Subalpine vegetation of the Wasatch Plateau, Utah. Ecol. Monog. 24:89-184.

58. Engelmann, G. 1880. Revision of the genus *Pinus,* and description of *Pinus Elliotti.* Trans. St. Louis Acad. Sci. 4:161-190.

59. Erdtman, G. 1943. *An Introduction to Pollen Analysis.* Ronald Press Co., New York. 239 pp.

60. ——— 1957. *Pollen and Spore Morphology, Plant Taxonomy.* Ronald Press Co., New York. 151 pp.

61. Faegri, K. and J. Iversen. 1950. *Textbook of Modern Pollen Analysis.* E. Munksgaard, Copenhagen. 168 pp.

62. Flint, R. F. 1957. *Glacial and Pleistocene Geology.* John Wiley & Sons, New York. 553 pp.

63. ——— and W. A. Gale. 1958. Stratigraphy and radiocarbon dates at Searles Lake, California. Amer. Jour. Sci. 256:689-714.

64. Frey, D. G. 1953. Regional aspects of the late-glacial and post-glacial pollen succession of southeastern North Carolina. Ecol. Monog. 23:289-313.

65. Gardner, J. L. 1951. Vegetation of the creosote bush area of the Rio Grande Valley in New Mexico. Ecol. Monog. 21:379-403.

66. Garner, H. F. 1959. Stratigraphic-sedimentary significance of contemporary climate and relief in four regions of the Andes Mountains. Bull. Geol. Soc. Amer. 70:1327-1368.

67. Gentry, H. S. 1957. *Los Pastizales de Durango.* Inst. Mexicano de Recursos Naturales Renovables, Mexico, D. F. 361 pp.

68. Gilluly, J. 1956. General geology of central Cochise County, Arizona. Geol. Surv. Prof. Paper 281:1-169.

69. Gladwin, H. S. 1947. Tree-rings and droughts. Medallion Papers 37:1-36.

70. Glendening, G. E. 1952. Some quantitative data on the increase of mesquite and cactus on a desert grassland range in southern Arizona. Ecology 33:319-328.

71. ——— and H. A. Paulsen. 1955. Reproduction and establishment of velvet mesquite as related to invasion of semi-desert grasslands. Tech. Bull. No. 1127. U. S. Dept. Agr., Forest Service. 50 pp.

72. Graham, A. and C. Heimsch. 1960. Pollen studies of some Texas peat deposits. Ecology 41:751-763.

73. Gray, J. 1960. Micropaleobotanical research on the late Tertiary sediments of Arizona. Ariz. Geol. Soc. Digest 3:145-149.

74. ——— 1961. Early Pleistocene paleoclimatic record from Sonoran Desert, Arizona. Science 113:38-39.

75. Hafsten, U. 1961. Pleistocene development of vegetation and climate as evidenced by pollen analysis. In *Contributions to the study of late Pleistocene environments of the southern high plains,* F. Wendorf, ed. Museum of New Mexico.

76. Haury, E. W. 1957. An alluvial site on the San Carlos Indian Reservation, Arizona. Amer. Antiquity 23:2-27.

77. ——— 1958a. Evidence at Point of Pines for a prehistoric migration from northern Arizona. In *Migrations in New World Culture History,* R. H. Thompson, ed. Univ. Ariz. Social Sci. Bull. No. 27.

78. ——— 1958b. Post-Pleistocene human occupation of the Southwest. In *Climate and Man in the Southwest,* T. L. Smiley, ed. Univ. Ariz. Bull. 28 (4): 69-73.

79. ——— 1960. Association of fossil fauna and artifacts of the Sulphur Spring stage, Cochise culture. Amer. Antiquity 25:609-610.

80. ———, E. B. Sayles, and William W. Wasley. 1959. The Lehner mammoth site, southeastern Arizona. Amer. Antiquity 25 (1):2-30.

81. Heindl, L. A. 1958. Cenozoic alluvial deposits of the upper Gila River area, New Mexico and Arizona. Univ. Arizona Ph.D. thesis.

82. Hernandez, E. and M. H. Gonzalez. 1959. Los pastizales de Chihuahua. Sec. Agricultura y Ganadería, Circular la Campana No. 3:1-48.

83. Hester, J. J. 1960. Late Pleistocene extinction and radiocarbon dating. Amer. Antiquity 26:58-77.

84. Hevly, R. and P. S. Martin. 1961. The geochronology of pluvial Lake Cochise. I. Pollen analysis of shore deposits. Jour. Ariz. Acad. Sci. 2:24-31.

85. Hill, R. T. 1896. Descriptive topographic terms of Spanish America. National Geographic 7:291-302.

86. Hoffmeister, D. F. and W. W. Goodpaster. 1954. The mammals of the Huachuca Mountains, southeastern Arizona. Illinois Biol. Monog. 24, 152 pp.

87. Hubbs, C. L. and C. Hubbs. 1953. An improved graphical analysis and comparison of series of samples. Systematic Zoology 2:49-57.

88. Huffington, R. M. and C. C. Albritton. 1941. Quaternary sands on the southern high plains of western Texas. Amer. Jour. Sci. 239:324-338.

89. Humphrey, R. R. 1958. The desert grassland. Botanical Review 24:193-252.

90. —— and L. A. Mehrhoff. 1958. Vegetation changes on a southern Arizona grassland range. Ecology 39:720-726.

91. Hunt, C. B. 1953. Pleistocene-recent boundary in the Rocky Mountain region. U. S. Geol. Survey Bull. 996-A. 25 pp.

92. Huntington, E. 1914. The climatic factor as illustrated in arid America. Carnegie Inst. Wash. Publ. 192. 341 pp.

93. Hutchinson, G. P., R. Patrick, E. S. Deevey. 1956. Sediments of Lake Patzcuaro, Michoacan, Mexico. Bull. Geol. Soc. Amer. 67:1491-1504.

94. Jones, M. D. and L. C. Newell. 1948. Size, variability, and identification of grass pollen. Jour. Amer. Soc. Agron. 40:136-143.

95. Judson, S. 1953. Geology of the San Jon site, eastern New Mexico. Smithsonian Misc. Coll. 121. 70 pp.

96. Kearney, T. H. and R. H. Peebles. 1951. *Arizona Flora.* Univ. of Calif. Press, Berkeley.

97. Keppel, R. V., J. E. Fletcher, J. L. Gardner, and K. G. Renard. 1958, 1959, 1960. Southwest Watershed Hydrology Studies Group, Annual Progress Reports. Tucson.

98. Knechtel, M. M. 1936. Geologic relations of the Gila conglomerate in southeastern Arizona. Amer. Jour. Sci. 31:81-92.

99. Lance, J. F. 1959. Geologic framework of arid basins in Arizona. Bull. Geol. Soc. Amer. 70:1729-1730.

100. —— 1960. Stratigraphic and structural position of Cenozoic fossil localities in Arizona. Ariz. Geol. Soc. Digest 3:155-159.

101. Laudermilk, J. D. and P. A. Munz. 1934. Plants in the dung of *Nothrotherium* from Gypsum Cave, Nevada. Carn. Inst. Wash. Publ. 453:31-37.

102. —— 1938. Plants in the dung of *Nothrotherium* from Rampart and Mauv Caves, Arizona. Carn. Inst. Wash. Publ. 487:273-281.

103. Leopold, A. S. 1950. Vegetation zones of Mexico. Ecology 31:507-518.

104. Leopold, L. B. 1951. Rainfall frequency: an aspect of climatic variation. Trans. Amer. Geophys. Union 32:347-357.

105. —— and J. P. Miller. 1954. A postglacial chronology for some alluvial valleys in Wyoming. Geol. Surv. Water-Supply Paper 1261:1-87.

106. Lindsay, A. J., Jr. 1958. Fossil pollen and its bearing on the archaeology of the Lehner mammoth site. Univ. Ariz. M.S. thesis, unpubl.

107. Linsley, E. G. 1958. The ecology of solitary bees. Hilgardia 27 (19):543-599.

108. Little, E. L. Jr. and R. S. Campbell. 1943. Flora of Jornada Experimental Range, New Mexico. Amer. Midland Nat. 30 (3):626-670.

109. Lowe, C. H. 1955. The eastern limit of the Sonoran desert in the United States with additions to the known herpetofauna of New Mexico. Ecology 36:343-345.

110. —— 1961. Biotic communities in the sub-Mogollon region of the inland Southwest. Jour. Ariz. Acad. Sci. 2:40-49.

111. MacDougal, D. T. 1908. Botanical features of North American deserts. Carn. Inst. Wash. Publ. 99:1-111.

112. Maher, L. J. 1961. Pollen analysis and postglacial vegetation history in the Animas Valley region, southern San Juan Mountains, Colorado. Ph.D. thesis, Univ. Minnesota.

113. Marr, John W. 1961. Ecosystems of the slope of the Front Range in Colorado. Univ. Colorado Press. 134 pp.

114. Marshall, J. T., Jr. 1956. Summer birds of the Rincon Mountains, Saguaro National Monument, Arizona. Condor 58:81-97.

115. —— 1957. Birds of pine-oak woodland in southern Arizona and adjacent Mexico. Pacific Coast Avifauna 32. 125 pp.

116. Martin, P .S. 1958a. A biogeography of reptiles and amphibians in the Gómez Farías region, Tamaulipas, Mexico. Misc. Publ. Mus. Zool. Univ. Mich. 101:1-102.

117. —— 1958b. Taiga-tundra and the full-glacial period in Chester County, Pennsylvania. Amer. Jour. Sci. 256:470-502.

118. —— 1958c. Pleistocene ecology and biogeography of North America. In *Zoogeography,* C. Hubbs, ed., Amer. Assoc. Advancement of Science Publ. 51:375-420.

119. —— 1961. Southwestern animal communities in the late Pleistocene. In *Bioecology of the Arid and Semiarid Lands of the Southwest,* L. M. Shields and L. J. Gardner, eds. New Mexico Highlands Univ. Bull., pp. 56-66.

120. ——, C. R. Robins and W. B. Heed. 1954. Birds and biogeography of the Sierra de Tamaulipas, an isolated pine-oak habitat. Wilson Bull., 66:38-57.

121. —— and J. Schoenwetter. 1960. Arizona's oldest cornfield. Science 132:33-34.

122. ——, B. Sabels and D. Shutler. 1961. Rampart Cave coprolite and ecology of the Shasta ground sloth. Amer. Jour. Sci. 259:102-127.

123. ——, J. Schoenwetter, and B. C. Arms. 1961. Palynology and Prehistory: The last 10,000 years. Geochronology Labs., Univ. Arizona (processed).

124. —— and J. Gray. 1962. Pollen analysis and the Cenozoic. Science 137:103-111.

125. Martin, P. S.,* J. B. Rinaldo, W. A. Longacre, C. Cronin, L. G. Freeman, Jr., and J. Schoenwetter. 1962. Chapters in the pre-history of eastern Arizona, I. *Fieldiana:* Anthropology 53:1-244.

126. Mason, R. J. 1962. The paleo-Indian tradition in eastern North America. Current Anthropology 3:227-278.

127. McDonald, J. E. 1956. Variability of precipitation in an arid region: a survey of characteristics for Arizona. Univ. Ariz. Inst. Atmospheric Physics Tech. Rep. 1:1-88.

128. McGrew, P. C. 1948. The Blancan faunas, their age and correlation. Bull. Geol. Soc. Amer. 59:549-552.

129. Meier, F. C. and E. Artschwager. 1938. Airplane collections of sugar-beet pollen. Science 88:507-508.

130. Meinzer, O. E. 1927. Plants as indicators of ground water. U. S. Geol. Surv. Water-supply Paper 577:1-95.

131. —— and F. C. Kelton. 1913. Geology and water resources of the Sulphur Spring Valley, Arizona. U. S. Geol. Surv. Water-supply Paper 320:9-213.

132. Melton, F. A. 1940. A tentative classification of sand dunes. Jour. Geol. 48:113-145.

133. Melton, M. A. 1960. Origin of the drainage of southeastern Arizona. Ariz. Geol. Soc. Digest 3:113-122.

134. Merriam, C. H. 1890. Results of a biological survey of the San Francisco Mountain region and desert of the Little Colorado in Arizona. North Amer. Fauna 3. Dept. Agr., Wash. 128 pp.

135. Miller, J. P. 1958. Problems of the Pleistocene in Cordilleran North America, as related to reconstruction of environmental changes that affected Early Man. In *Climate and Man in the Southwest,* T. L. Smiley, ed. Univ. Ariz. Bull. 28 (4):19-49.

136. —— and F. Wendorf. 1958. Alluvial chronology of the Tesuque Valley, New Mexico. Jour. Geol. 66:177-194.

137. Morrison, R. B., J. Gilluly, G. M. Richmond, and C. B. Hunt. 1957. In behalf of the Recent. Amer. Jour. Sci. 255:385-393.

138. Muller, C. H. 1947. Vegetation and climate of Coahuila, Mexico. Madroño 9:33-57.

139. Muller, J. 1959. Palynology of recent Orinoco delta and shelf sediments: reports of the Orinoco Shelf Expedition. Micropaleontology 5:1-32.

140. Murray, G. W. 1951. The Egyptian climate: an historical outline. Geog. Jour. 117:422-434.

141. Nelson, E. W. 1934. The influence of precipitation and grazing upon black grama grass range. Tech. Bull. 409, U. S. Dept. Agr. 32 pp.

---

*Curator of Anthropology, Chicago Natural History Museum, and a leading contributor to archaeological studies of the Southwest. Although unrelated, we share the same name.

142. Nichol, A. A. 1952. The natural vegetation of Arizona. Univ. Ariz., College Agr. Techn. Bull. 127:189-230.

143. Parker, K. W. and S. C. Martin. 1952. The mesquite problem on southern Arizona ranges. U. S. Dept. Agr. Cir. No. 908. 70 pp.

144. Potter, L. D. and J. Rowley. 1960. Pollen rain and vegetation, San Augustin Plains, New Mexico. Bot. Gazette 122:1-25.

145. Preston, F. W. 1960. Time and space and the variation of species. Ecology 41:611-627.

146. Quinn, J. H. 1957. Paired river terraces and Pleistocene glaciation. Jour. Geol. 65:149-166.

147. Roosma, A. 1958. A climatic record from Searles Lake, California. Science 128:716.

148. Rothrock, J. T. 1878. Reports upon the botanical collections. In *Report upon United States Geographical Surveys West of the One Hundredth Meridian*, Vol. VI, Govt. Print. Office, Wash.

149. Russell, I. C. 1885. *Geological History of Lake Lahontan.* U. S. Geological Survey Monographs 11:1-288.

150. Sabins, F. F. 1957. Geology of the Cochise Head and western part of the Vanar quadrangles, Arizona. Geol. Soc. Amer. Bull. 68:1315-1342.

151. Sayles, E. B. and E. Antevs. 1941. The Cochise culture. Medallion Papers 29:1-81.

152. ——, et al. MS. The Cochise culture.

153. Schoenwetter, J. 1960. Pollen analysis of sediments from Matty Wash. Univ. Ariz. M.S. thesis, 72 pp.

154. Schott, A. 1857. Geological observations. In *Report on the United States and Mexican Boundary Survey*, Vol. II, W. H.·Emory. U. S. Geol. Survey, Gov. Print. Office, Wash.

155. Schulman, E. 1956. Dendroclimatic changes in semiarid America. Univ. Ariz. Press. 142 pp.

156. Schumm, S. A. and R. F. Hadley. 1957. Arroyos and the semi-arid cycle of erosion. Am. Jour. Sci. 255:161-174.

157. Sears, P. B. 1937. Pollen analysis as an aid in dating cultural deposits in the United States. In *Early Man*, G. G. MacCurdy, ed. Lippincott Co., London, pp. 61-66.

158. ——, F. Foreman, K. H. Clisby. 1955. Palynology in southern North America. Bull. Geol. Soc. Amer. 66:471-530.

159. ——, and A. Roosma. 1961. A climatic sequence from two Nevada caves. Amer. Jour. Sci. 259:669-678.

160. Sellers, W. D. 1960. Precipitation trends in Arizona and western New Mexico. In *Proceedings, 28th Annual Western Snow Conference*, Santa Fe, N. M., pp. 81-94.

161. Selling, O. H. 1948. Studies in Hawaiian pollen statistics Part III. On the late Quaternary history of the Hawaiian vegetation. Bernice P. Bishop Mus. Special Pub. 39:1-154.

162. Shreve, F. 1915. The vegetation of a desert mountain range as conditioned by climatic factors. Carn. Inst. Wash. Pub. 217:1-112.

163. —— 1919. Comparison of the vegetational features of two desert mountain ranges. Plant World 22:291-307.

164. —— 1922. Conditions indirectly affecting vertical distribution on desert mountains. Ecology 3:269-274.

165. —— 1939. Observations on the vegetation of Chihuahua. Madroño 5:1-13.

166. —— 1942a. Grassland and related vegetation in northern Mexico. Madroño 6:190-198.

167. —— 1942b. The desert vegetation of North America. Bot. Rev. 8:195-246.

168. —— 1944. Rainfall of northern Mexico. Ecology 25:105-111.

169. —— 1951. Vegetation of the Sonoran Desert. Carn. Inst. Wash. Pub. 591. 192 pp.

170. Steenis, C. G. G. J. Van. 1934—1935. On the origin of the Malaysian mountain flora. Bull. Jard. Bot. Buitenzorg, Bull. Ser. 3, 13:135-262; 289-417.

171. Steeves, M. W. and E. S. Barghoorn. 1959. The pollen of *Ephedra.* Jour. Arnold Arboretum 40:221-255.

172. Thornber, J. J. 1910. The grazing ranges of Arizona. Ariz. Exp. Sta. Bull. 65:245-360.

173. Topia, C. and J. de Alba. 1955. Species survey of a Mexican unfenced range. Jour. Range Management 8:111-114.

174. Tuan, Yi-Fu, 1959. Pediments in southeastern Arizona. Univ. Calif. Publications in Geog. 13:1-140.

175. Walkington, D. L. 1960. A survey of the hay fever plants and important atmospheric allergens in the Phoenix, Arizona metropolitan area. Jour. Allergy 31:25-41.

176. Wallen, C. C. 1955. Some characteristics of precipitation in Mexico. Geografiska Annaler 37:51-85.

177. Wallmo, O. C. 1955. Vegetation of the Huachuca Mountains, Arizona. Amer. Midland Nat. 54:466-480.

178. Wenner, C. G. 1947. Pollen diagrams from Labrador. Geografiska Annaler 29:137-373.

179. White, Stephen S. 1949. The vegetation and flora of the region of the Río de Bavispe in northeastern Sonora, Mexico. Lloydia 11:229-302.

180. Whittaker, R. H. 1956. Vegetation of the Great Smoky Mountains. Ecol. Monog. 26:1-80.

181. Wise, E. N. and D. Shutler. 1958. University of Arizona radiocarbon dates. Science 127:72-74.

182. Wodehouse, R. P. 1935. *Pollen Grains.* McGraw-Hill Co., New York. 574 pp.

183. Wood, P. A. 1960. Paleontological investigations in the 111 Ranch area. Ariz. Geol. Soc. Digest 3:141-143.

184. Woodbury, A. M. 1947. Distribution of pigmy conifers in Utah and northeastern Arizona. Ecology 28:113-126.

185. Wright, H. E. 1952. The geological setting of four prehistoric sites in northeastern Iraq. Bull. American Schools of Oriental Research 128:11-24.

186. Yang, T. W. and C. H. Lowe. 1956. Correlation of major vegetation climaxes with soil characteristics in the Sonoran Desert. Science 123:542.

1. *Upper edge of the desert grassland, Empire Valley, 1,500 m.;* Quercus emoryi *occupies the north-, grama grassland the south-facing slope; Spikes of* Muhlenbergia *in foreground. August 16, 1958.*

2. *Cienega of the Empire Wash, 14 km. north northeast of Sonoita. This undissected flood plain is dominated by* Sporobolus; *mesquite in background.*

3. Cattle tank on Cienega Ranch, Empire Valley, August 2, 1958.
Note bunches of Sporobolus *and seedling* Amaranthus.

4. Same, August 16, 1958. Emergent Amaranthus *now hides the* Sporobolus.
This is the summer season of maximum growth.

5. *Vegetation of the Sulphur Spring Valley, encinal with* Pinus cembroides, Quercus hypoleucoides, Q. toumeyi, Q. arizonica, Arctostaphylos *and* Yucca, *top of Mule Mountains, 2,100 m., November 1959.*

6. *Vegetation of the Sulphur Spring Valley, plotless sample station two near Dixie Ranch;* Fouquieria splendens, Calliandra *and various grasses.*

7. *Vegetation of the Sulphur Spring Valley, mesquite mounds near plotless sampling station ten. Dead whitish spikes of* Amaranthus palmeri *projecting through the leafless mesquite* (Prosopis juliflora). *The hard sandy flats between the mounds are barren even in the rainy season. November 1959.*

8. Vegetation of the Sulphur Spring Valley, dry pond with crop of Xanthium *in fruit; desert grassland* (Hilaria, Bouteloua, *forbs*) *in background.*

9. Vegetation of the Sulphur Spring Valley, Whitewater Draw at El Paso gas pipeline crossing. The Whitewater Draw is beginning to fill; arroyo banks cut in the early 20th century have slumped and are scarcely visible. Mesquite, Baccharis glutinosus, Sporobolus, Johnson Grass. November 1959.

11. Pollen Profile 1, Cienega Creek, Empire Valley.

10. Vegetation of the Sulphur Spring Valley, Yucca elata 4.8 m. tall, in desert grassland one km. east of McNeal generating plant and three km. south of plotless sampling station 14.

12. Pollen Profile 3, Double Adobe I.

13. Pollen Profile 3, Double Adobe I.

14. Pollen Profile 6, Malpais site, Chihuahua.

INDEX

Alluviation: 2-3, 36, 47, 61-62, 67-69
Alluvium: 1-3, 15, 28, 34, 36, 38, 44, 48-50, 64-65, 69
Alluvial Chronology: 61, 64
Altithermal period: 34, 36, 39, 42, 49, 50, 60-61, 64-65, 67-68, 70
Amaranthus see Cheno-ams
Arroyo cutting: 3, 49, 60-61, 64, 66-67, 69
Atlantic High: 4

Bajadas: 1, 2, 7, 14, 23, 31, 44, 68, 69
Basin and range province: 1
Bedrock, effect on vegetation: 11, 49
Biochores: 5 ff.

Caliche formation: 60, 65
Carbon, organic: 56-57
Cazador type site, fossil pollens: 30, 38-39, 40, 60
Cheno-am pollens: 21, 23, 34, 36, 38, 39, 42, 44, 49, 58, 59, 69
Chenopodiaceae *see* Cheno-ams
Chiricahua Mountains and Merriam effect: 9
Chiracahua stage, fossil pollen: 39, 41, 48, 60
Cienega Creek, Empire Valley, fossil pollen: 30, 31, 32, 34, 36, 49-50, 59
Cienega Creek, Point of Pines, fossil pollen: 30, 34, 48, 49, 50, 70
Cienegas: 3, 34, 42, 44, 59, 60, 69
Climate during alluviation: 2, 47-48, 61, 64, 67
Climate, postpluvial: 59 ff., 67
    see also Altithermal period, Drought
Cochise Culture: 30, 36, 48, 60, 69
Compositae pollen: 21, 23, 34, 36, 38, 39, 42, 44, 47, 49, 56, 58, 59, 69
Coniferous forest: 5-6
Corn *see Zea*
Cutting and filling: 3, 34, 39, 48, 59, 69

Desert biochore: 7-8
Desert grassland: 7, 13 ff., 42, 44, 47, 48, 49, 52, 58, 59, 60, 61, 67
Desert grassland pollen zones: 31 ff., 47, 59, 68, 69
Double Adobe: 29, 30, 34-41, 42, 49, 52, 53, 54, 56, 59, 60
Drought period: 42, 66-67, 70
Dry Prong Reservoir, fossil pollen: 47-49, 53-53, 61
Dunes: 65-66

Early Man: 44, 70
Edaphic conditions: 11
Elevation, effect on vegetation: 8
Empire Cienega: 21, 23, 34, 47, 59
Encinal: 6-7, 67, 68
Environment and vegetation: 8 ff.
Extinction: 39, 44, 47, 61, 64-65, 70

Fire, effect on vegetation: 13, 68
Flood plains: vi-vii, 23, 31, 34, 42, 44, 47, 48, 49, 55, 58, 59, 68-70
Forest biochore: 5-6

Grass pollen: 23, 28, 34, 36, 38, 39, 48, 49, 59, 60
Grassland biochore: 7
Grassland *see* Desert grassland
Great Drought: 66-67, 70

Huachuca Mountains and Merriam effect: 9

Inner valley fill: 2-3
Insects and pollen: 28
Inversions, effect on vegetation: 11-12

Laboratory methods: 30
Latitude, effect on vegetation: 8-9, 67
Lehner site, fossil pollen: 30, 44, 47, 59, 60, 69
Life zones: 8, 66

Malpais site, fossil pollen: 30, 39, 42, 49, 60
Matty Wash, fossil pollen: 30, 33, 34, 49, 59
Megafauna, extinction of, *see* Extinction
Mesquite pollen: 21, 28, 38, 52
Merriam effect: 9-11
Mexican plants and animals in southern Arizona: vi, 68, 70
Monsoon: 3-4, 61, 67, 68
Mountain sides and vegetation: 9
Murray Springs, fossil pollen: 30, 39, 42, 43, 49, 50, 53-54, 59, 61

Oak: 6-7
Oak pollen: 21, 34, 39, 60, 69

Palynology: v, 42
Pine-oak woodland: 6-7, 66-67
Pine pollen: 15, 19, 21, 34, 36, 38, 39, 42, 47, 48, 49, 52, 53, 54, 58, 59, 60, 61, 66, 67, 69
Pollen: v
Pollen analysis: 5, 30-31, 47, 68, 69
Pollen analysis and archeology: 28 ff, 42, 44, 47, 48, 68, 70
Pollen extraction and counting: 30, 31
Pollen fossil:
    Abies, Picea and *Alnus*: 47, 54-55, 59, 60, 66
    Altithermal: 60-61
    Artemesia: 38, 47, 49, 52, 59, 60, 66
    Betula: 39, 55-56
    Carya-Ulmus: 36, 38, 39, 48, 49, 54-55, 56, 59, 60
    Celtis: 56
    Cheno-ams: 34, 36, 38, 39, 42, 44, 48, 49, 58, 59, 69
    Compositae: 34, 36, 38, 39, 42, 44, 47, 48, 49, 56, 58, 59, 69
    Cyperaceae: 34, 36, 38, 39, 42, 47, 48, 50, 60
    Ephedra: 42, 48, 51-52, 61
    Eriogonum: 49
    Euphorbiaceae: 48, 49, 50
    Fraxinus: 36, 38, 44, 54, 55, 59
    Juglans: 55
    Juniperus: 39, 60
    Kallstroemia: 36, 49, 60
    Liguliflorae: 38, 42, 59
    Malvaceae: 36, 49, 50

86